THINKING ABOUT BIOLOGY

AN INTRODUCTORY LABORATORY MANUAL
FOURTH EDITION

MIMI BRES • ARNOLD WEISSHAAR
PRINCE GEORGE'S COMMUNITY COLLEGE

PEARSON

Boston   Columbus   Indianapolis   New York   San Francisco   Upper Saddle River
Amsterdam   Cape Town   Dubai   London   Madrid   Milan   Munich   Paris   Montreal   Toronto
Delhi   Mexico City   Sao Paulo   Sydney   Hong Kong   Seoul   Singapore   Taipei   Tokyo

Editor-in-Chief: Beth Wilbur
Senior Acquisitions Editor: Star
   MacKenzie
Assistant Editor: Frances Sink
Executive Director of Development:
   Deborah Gale
Marketing Manager: Lauren Rodgers
Director of Production: Erin Gregg
Managing Editor: Michael Early
Production Project Manager: Lori
   Newman

Production Management: Laserwords
Copyeditor: Linda Benson
Compositor: Laserwords
Cover Designer: Central Design
Illustrators: Laserwords
Photo Researcher: Bill Smith Group
Image Lead: Donna Kalal
Manufacturing Buyer: Michael Penne
Cover & Interior Printer: Edwards
   Brothers
Cover Photo Credit: Photolibrary

Credits and acknowledgments borrowed from other sources and reproduced, with permission, in this textbook appear on pages 445–446.

Many of the designations used by manufacturers and sellers to distinguish their products are claimed as trademarks. Where those designations appear in this book, and the publisher was aware of a trademark claim, the designations have been printed in initial caps or all caps.

Benjamin Cummings is a trademark, in the U.S. and/or other countries, of Pearson Education, Inc. or its affiliates.

Library of Congress Cataloging-in-Publication Data

Bres, Mimi.
   Thinking about biology : an introductory laboratory manual / Mimi Bres, Arnold Weisshaar. — 4th ed.
     p. cm.

  1. Biology—Laboratory manuals.   I. Weisshaar, Arnold.   II. Title.
   QH317.B7935 2013
   570.78—dc23

2011041178

1 2 3 4 5 6 7 8 9 10—EDB—15 14 13 12 11

PEARSON

www.pearsonhighered.com

ISBN 10: 0-321-79195-9
ISBN 13: 978-0-321-79195-5

Contents

Contents

Preface

Thinking About Biology is designed for a one-semester, general biology course for non-majors. The topics and exercises are general enough to be compatible with any introductory-level human and general biology text. The activities demonstrate that basic biological concepts can be applied to a wide variety of plants, animals, and microorganisms.

This book is unique not because of the specific topics covered, but because of the approach to these topics. The laboratory exercises are planned to help you

- gain practical experience that will help you understand lecture concepts
- acquire the basic knowledge needed to make informed decisions about biological questions that arise in everyday life
- develop the problem-solving skills that will lead to success in college and in a competitive job market
- learn to work effectively and productively as a member of a team

The basis of scientific work is asking questions and answering them by observations or experiments. Thus, we strongly believe that the most important goal of an introductory course in the life sciences is to achieve some understanding of the processes of investigation that are basic to science, and how scientists work to solve problems—not to simply memorize facts about animals, molecules, or new techniques.

We hope that working through this laboratory manual will be an exciting experience for both students and instructors, and one that will leave you better prepared to meet the demands of our increasingly scientific society.

Wishing you an enjoyable and successful semester.

Mimi Bres, Ph.D.
Professor of Biological Sciences
Prince George's Community College
Largo, Maryland

Arnold Weisshaar, M.S. Physiology
Professor Emeritus of Biological Sciences
Prince George's Community College
Largo, Maryland

NEW FOR THE FOURTH EDITION

New and revised activities for this edition were based on feedback from hundreds of students and a selection of faculty reviewers from around the United States. Among the key features, you'll find

- a reorganized table of contents
- new activities on cells, photosynthesis, and ecosystems
- new evolution exercise
- more photos and diagrams
- cutting-edge activities on forensic DNA analysis
- revised assessment questions

SPECIAL FEATURES FOR STUDENTS

Informal Style

- The text is written simply with easy-to-understand language. Key terms and important definitions are highlighted with bold print for easy recognition. Large spaces are provided throughout the manual for you to record and fully explain your answers.

Active Learning Experience

- Every exercise gives you an opportunity to be an investigator using the scientific method. You'll form hypotheses, set up experiments, collect data, record your data in graphs and charts, and draw conclusions from your experimental results.

Team-Building Opportunities

- Most laboratory activities emphasize a team approach. Group work is encouraged and often required. In the real-world job market, you'll be expected to interact with others to solve problems and complete projects. This approach provides opportunities for you to work together, share ideas, and function effectively in groups to accomplish tasks.

Real-Life Connections

- Lab activities are designed to stimulate interest in topics that can help you make decisions regarding your own health and nutrition, understand current topics in the news, and become informed about how your personal actions affect the environment.

Tools for Success

The following components of each exercise will help you succeed:

- **Instructional Objectives.** The objectives are listed first in each exercise so that you'll be able to focus your attention on the main concepts of each activity.

- **Content Focus.** Each exercise includes a brief discussion of the background information that you'll need to understand the subject of the exercise and to prepare you to complete the activities that follow.
- **Notes and Cautions.** Note boxes provide helpful hints for solving problems and accomplishing laboratory tasks. Pay special attention to caution boxes that provide important safety information.
- **Comprehension Checks.** Stop and complete the Comprehension Check questions to get immediate feedback on your understanding of the basic principles covered in each activity. These questions also provide a chance for you to apply what you've learned to situations outside the classroom.
- **Check-off Boxes.** These boxes allow your instructor to check your progress and make sure you've understood the important concepts in each activity. This will help you focus your study time and be more successful in the course.
- **Self Tests.** Answer these questions after completing the laboratory exercise. The Self Test questions allow you to assess your comprehension and apply your knowledge. You can find the answers to the Self Test questions in Appendix II at the back of the book.

ACKNOWLEDGMENTS

A project of this size and scope can't be completed without support and cooperation from many people. We would like to express our deep appreciation for their contributions.

Thanks to

Christine Barrow, Dean of Sciences, Technology, Engineering, and Mathematics at Prince George's Community College, for providing the opportunity to do this project and supporting professional development.

Cassandra Moore-Crawford, professor, Department of Biological Sciences, Prince George's Community College, for enabling us to think "outside the box" when developing these lab activities.

Michael Garvey, deputy managing director of the Philadelphia Police Department, director of Forensic Services, for providing your time and technical expertise in the area of forensic molecular genetics.

Kenneth Thomulka, assistant professor of biology, University of the Sciences in Philadelphia, for allowing us to adapt an imaginative and unique application of biotechnology to the education arena.

John Kasmer, Ph.D., associate professor of biology, Northeastern Illinois University, for the development of the methods and procedures for the brine shrimp "mark and recapture" activity.

The Biology 1010 faculty at Prince George's Community College. Your many helpful suggestions have contributed substantially to the effectiveness of these laboratory exercises.

The Biology 1010 students for classroom-testing our laboratory exercises and offering suggestions for improvements.

Frances Sink, assistant editor, Pearson Higher Education, for lending your skill and support to this project. Your efficient and effective management played a significant role in making this fourth edition possible.

Karen Berry, production editor, Laserwords, for her help and timely responses to issues that arose during the coordination of the art, text, editing, and other details necessary to the production of this manual.

The reviewers listed below graciously gave their time to comment on this new edition and make useful suggestions. For their help, many thanks.

REVIEWERS

Dale Amos, *University of Arkansas–Fort Smith*

Cheryl Boice, *Lake City Community College*

John S. Campbell, *Northwest College*

Reggie Cobb, *Nash Community College*

Beverly Cochran, *Texas A&M University–Commerce*

Sandra Gibbons, *Moraine Valley Community College*

Aimee Howe, *McLennan Community College*

Suzanne Long, *Monroe Community College*

Jameson McCann, *Guilford Technical Community College*

Aileen Miller-Jenkins, *Edward Waters College*

Sharon E. Mozley-Standridge, *Middle Georgia College*

Robin Patterson, *Butler County Community College*

Virginia Rivers, *Truckee Meadows Community College*

Introduction to the Scientific Method

Objectives

After completing this exercise, you should be able to:

- use the scientific method to solve problems
- organize information to facilitate analysis of your data
- draw graphs that present data clearly and accurately
- interpret data in tables, charts, and graphs
- draw conclusions that are supported by experimental data
- analyze data using common statistical measures
- apply your knowledge of the scientific method to real-life situations

CONTENT FOCUS

What is science? What do scientists *do* all day? For most of you, these aren't easy questions to answer. The widespread picture of a middle-aged man in a white lab coat doesn't apply to most scientists. So, what are scientists really like? They all have the "**four Cs**" in common.

Just like you, scientists are **curious** about the world around them. They ask questions about everything. Can my diet cause heart disease? Why does the river look brown instead of blue? How can squirrels remember where they bury their nuts? Why do some cars get better mileage than others? Science is a method for answering these and many other questions.

Scientists don't accept things without **collecting information.** All the facts relating to a problem or question have to be carefully explored and checked for accuracy. The most common way to collect information is through controlled experiments. Answering the complicated questions asked by today's scientists may require many different experiments performed over months or even years.

1

When scientists complete a series of experiments, they **communicate the results** of their research to their peers and the scientific community by publishing papers in print or Internet journals.

When the research results are made public, subject matter experts from around the world will do similar experiments to duplicate and verify the published results. Extensive verification is needed before experimental results are accepted as fact by the scientific community.

Scientists are **comfortable with new concepts.** If a better explanation can be found, scientists aren't afraid to give up old ideas for new ones.

To make the four Cs happen, scientists have developed a series of steps in investigation called **the scientific method.** Through trial and error, the scientific method has proved to be an efficient and effective way of attacking a problem. You've probably used some version of the scientific method many times in your life—without being aware of the steps you were following.

Note:

All the information, concepts, and relationships you'll read about in your textbook and laboratory manual were discovered and verified by the same scientific process you'll use in this exercise.

During this course, we'll be summarizing hundreds of years of study and experimentation. Information presented on television, in newspapers, and on the Internet often hasn't been confirmed by this same careful process and may not be correct.

ACTIVITY 1

FORMING HYPOTHESES TO SOLVE PROBLEMS

There are several ways that a problem can come to your attention. Someone may **assign** you the problem (this happens often in a school or a work situation), the problem may **thrust itself** upon you (your car won't start), or you may discover the problem by simply being **curious** about something you've seen. An easy way to attack the problem is to make an **educated guess** about the possible solution to the problem. It's an "educated" guess because you use all the background information that's available when making your guess. In scientific terms, an educated guess is called a **hypothesis.**

Let's begin with a simple situation that you might face in your college education.

> ### The Problem
> You were absent from chemistry class the day your professor gave out instructions to mix the chemicals needed for your laboratory experiment. Your roommate was in class and copied the instructions for you. You rush off to chemistry lab and mix the chemicals, but when you use them in your experiment, they don't perform as expected.

In **Table 1-1, list three hypotheses** about why the formula didn't work. Don't forget—hypotheses must be **testable!**

T A B L E 1 - 1 CHEMISTRY EXPERIMENT HYPOTHESES
A:
B:
C:

Check your hypotheses with your instructor before you continue.

Some hypotheses can be tested by observation only, but, more often, you'll need a **combination of observation and experimentation** to be sure about the accuracy of your results. To understand how scientists work, you must follow the steps of the scientific method as they are used in actual **experiments**. In Activities 2 and 3, you'll **see how scientific method skills** are used to **set up experiments** and analyze the **information (data)** that's collected.

ACTIVITY 2   TESTING HYPOTHESES

> ## The Problem
> **Investigate the effects of fertilizer on plant growth.**

STEP 1:

You form a hypothesis about what you think will happen.

Hypothesis: Adding fertilizer will make plants grow taller.

STEP 2:

You design an experiment that compares the growth (in height) of plants that **receive fertilizer** with those grown **without fertilizer. Your design might be similar to the following:**

Begin with **20 plants** of the same type and approximately the same size.

- An experiment is designed to isolate the factor **(variable)** you're interested in testing. Variables can change over time or under different experimental conditions. In a well-designed experiment, you should be able to easily **measure** the changes in your variables. For example, in this experiment, the changes in variable "plant height" will be measured.
    All other conditions must be held constant. In this way, you're sure that your observed results were caused by the only variable that was tested.
- Because you're investigating the **effect of fertilizer,** you'll want to hold all **other factors constant** (plant type, plant size, pot size, amount of water, amount of light, etc.) to avoid confusion. This way you can be **sure** that any differences in height are **due to the presence of fertilizer** and not to some other factor.

It's helpful to plan an experiment with a **group** of plants (or animals). There are **two good reasons** to use groups:

- If **unexpected factors** (such as death, disease, or accidents) affect a few experimental subjects, it won't ruin the experiment.
- **Natural genetic differences between individuals of the same species** will cause some plants to grow taller than others (just as some people grow taller than others). You can separate this effect from that of the fertilizer by measuring the height in a **group** of plants for each treatment (fertilizer and no fertilizer). Because there's so much variation among individuals, you can increase the validity and reliability of your results by testing **as large a group as possible.**

STEP 3:

You decide that **10 plants will receive identical, measured amounts of fertilizer** each week. These are the **experimental** plants. They are receiving the treatment (fertilizer) that will help you test your original hypothesis (fertilizer will make plants grow taller).

Ten other plants will receive no fertilizer. These are the **control** plants. They don't receive the experimental treatment. You'll use these for **comparison with the experimental group** to help you interpret your results and to show that any observed differences in height between the two groups are due to the **only difference between them**—the application of fertilizer.

In this example, there are **10 replications of the experimental treatment and 10 replications of the control treatment.**

You plan to measure the growth of your plants (height in centimeters) **once a week for a month.** You'll keep detailed **records** of your **observations.**

✔ Comprehension Check

1. Why is it necessary to divide the plants into two groups (a control group and an experimental group)?

2. Why is it important to keep conditions exactly the same in the control and experimental groups, **except** for the application of fertilizer?

3. If you were designing this experiment for a fertilizer manufacturer, what changes would you make in the experimental design to be confident that your results would be accepted as fact by the scientific community? **Explain** your answer.

Check your answers with your instructor before you continue.

ACTIVITY 3   INTERPRETING DATA

The month is up. You're ready to draw conclusions from your **data** (the information you have recorded). You'll be thinking about what your results mean and whether your hypothesis is **supported.** The information you collected during your experiment is presented in **Tables 1-2 and 1-3.**

TABLE 1-2
HEIGHT GAIN (cm) OVER FOUR WEEKS—CONTROL PLANTS

PLANT NUMBER	INITIAL HEIGHT (cm)	WEEK 1	WEEK 2	WEEK 3	WEEK 4	GROWTH OVER FOUR WEEKS (cm)
1	10.0	1.6	2.0	3.0	2.5	9.1
2	11.5	2.2	1.5	1.5	2.0	7.2
3	9.6	1.5	2.3	2.6	2.0	8.4
4	9.2	2.0	3.0	2.8	1.5	9.3
5	10.2	2.3	1.2	1.6	2.0	7.1
6	11.0	3.2	1.7	2.0	3.2	10.1
7	10.0	2.6	3.0	3.0	1.4	10.0
8	9.7	4.0	2.6	4.0	2.3	12.9
9	10.4	DIED	—	—	—	—
10	10.4	2.3	2.3	2.7	2.7	10.0
TOTAL HEIGHT GAIN—84.1 cm						
AVERAGE HEIGHT GAIN—9.3 cm						

TABLE 1-3

HEIGHT GAIN (cm) OVER FOUR WEEKS—EXPERIMENTAL PLANTS

PLANT NUMBER	INITIAL HEIGHT (cm)	WEEK 1	WEEK 2	WEEK 3	WEEK 4	GROWTH OVER FOUR WEEKS (cm)
1	9.6	4.2	5.0	3.0	4.7	16.9
2	9.8	6.0	4.0	5.5	5.0	20.5
3	10.3	5.3	5.5	3.6	4.2	18.6
4	11.0	2.1	3.2	6.2	3.8	15.3
5	10.1	3.4	4.0	4.4	4.0	15.8
6	9.2	4.7	3.1	3.1	4.0	14.9
7	9.5	4.2	5.2	3.9	3.6	16.9
8	10.0	3.3	6.0	5.6	4.2	19.1
9	9.7	5.8	6.1	6.5	5.0	23.4
10	10.4	5.1	3.4	5.8	5.3	19.6
TOTAL HEIGHT GAIN—181.0 cm						
AVERAGE HEIGHT GAIN—18.1 cm						

✔ Comprehension Check

1. Were there differences in growth between the control and experimental plants?

    If so, which group grew taller?

2. Do the results support the original hypothesis? **Explain** your answer.

3. Why is it more accurate to compare the **average** height gain of the control and experimental groups (instead of comparing individual plants)?

Check your answers with your instructor before you continue.

ACTIVITY 4

PERFORMING AN EXPERIMENT AND COLLECTING DATA

Regardless of where you live, you've probably heard a lot of news stories lately about the increasing spread of HIV (which causes AIDS) and other sexually transmitted infections throughout the United States. In this activity, you'll simulate how multiple sexual encounters by an infected person can lead to the spread of a disease through a population.

1. On your laboratory table, you'll find **a dropper bottle of distilled water** and **a dropper bottle of phenolphthalein solution** (a chemical indicator).

    From the supply area, get **four clean sampling cups, a marker pen,** and **a container that holds an unidentified liquid.**

> ### Note:
> The unidentified liquid in your container represents the body fluids that may be exchanged during sexual encounters. Your container holds either distilled water or a solution that changes color when exposed to phenolphthalein. A color change simulates body fluids that are infected with HIV. Only one container in the class is "infected."

2. Using a marker pen, label the first small sampling cup with the number 2 as shown in **Figure 1-1.** Label the others 4, 6, and D (which stands for distilled water).

FIGURE 1-1. Labeled Sampling Cups

3. Read through the following instructions on how to perform "body fluid" exchanges. **WHEN YOUR INSTRUCTOR TELLS YOU TO START, follow the steps listed** and complete **Exchange #1.**

   To "exchange" body fluids, follow these steps:
   - Randomly select **one partner** for the exchange.
   - Pour **all** of the solution in your container into your partner's container.
   - Your partner will pour **all** of the solution in his/her container back into your container.
   - Pour **only half** of the solution back into your partner's empty container.
   - Return to your seat. **Exchange #1** has been completed.

4. **WAIT** until all the students in the class have completed Exchange #1.

   WHEN YOUR INSTRUCTOR TELLS YOU TO START, perform **Exchange #2** using the same procedures as you did for Exchange #1.

   Don't select the same partner you had before.

5. Pour a small amount of liquid from your container of "body fluids" into the sampling cup labeled with the number 2 (just enough liquid to cover the bottom of your sampling cup).

   Set the sampling cup aside and wait for your instructor's signal to continue with the simulation.

6. **WHEN YOUR INSTRUCTOR TELLS YOU TO START,** perform **Exchange #3.**

   Don't select any partner you've had before.

7. **WHEN YOUR INSTRUCTOR TELLS YOU TO START,** perform **Exchange #4.**

   Don't select any partner you've had before.

8. Pour a small amount of liquid from your container of "body fluids" into the sampling cup labeled with the number 4 (just enough liquid to **cover the bottom** of your sampling cup).

   Set the sampling cup aside and wait for your instructor's signal to continue with the simulation.

9. **WHEN YOUR INSTRUCTOR TELLS YOU TO START,** perform **Exchange #5.**

   Don't select any partner you've had before.

10. **WHEN YOUR INSTRUCTOR TELLS YOU TO START,** perform **Exchange #6.**

    Don't select any partner you've had before.

11. Pour a small amount of liquid from your container of "body fluids" into the sampling cup labeled with the number 6 (just enough liquid to cover the bottom of your sampling cup).

12. Using the dropper bottle of distilled water, drop a small amount of **distilled water** into the sampling cup labeled with the letter D (just enough to cover the bottom of your sampling cup).

13. Place **one drop of phenolphthalein** indicator solution into each of the four sampling cups (cups 2, 4, 6, and D).

    If the liquid in any of the sampling cups **turns bright pink,** you've been **infected with HIV.**

    Based on the results of your indicator tests, check the appropriate box in **Table 1-4** (infected or not infected) for **Exchanges #2, #4, #6,** and the **distilled water.**

TABLE 1-4
RESULTS OF INDICATOR TESTS FOR BODY FLUID EXCHANGES

	INFECTED	NOT INFECTED
Exchange #2		
Exchange #4		
Exchange #6		
Distilled Water		

14. When everyone has recorded his/her results, your instructor will survey the class to determine the total number of infected students after **Exchanges #2, #4, and #6.** Record the results of the survey in **Table 1-5.**

T A B L E 1 - 5 SIMULATION RESULTS—SPREAD OF HIV	
EXPERIMENTAL TRIALS	NUMBER OF INFECTED CLASS MEMBERS
Start of experiment	1
Exchange #2	
Exchange #4	
Exchange #6	

15. What was the reason for performing the phenolphthalein indicator test on the distilled water? Explain your answer.

16. In scientific terminology, what name is given to the part of the simulation in which you performed the indicator test on the distilled water? _____

ACTIVITY 5   GRAPHING YOUR RESULTS

1. Graphs provide a good visual representation of the relationships between the factors investigated in an experiment.

2. Look at the graph structure in **Figure 1-2** and note the following **key points:**

   ■ The **horizontal** axis is referred to as the *X*-axis. The **vertical** axis is called the *Y*-axis.
   ■ Numbers on the X- and Y-axes must have an **equal interval** between them (for example, 5, 10, 15 but **not** 5, 10, 20, 50).

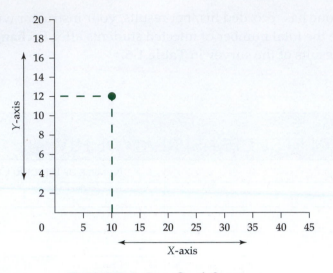

FIGURE 1-2.   Graph Structure

- When deciding which factors to plot on the X- and Y-axes, here's a good rule of thumb: if one data set consists of words (months of the year, car models, country names) and the other is numbers (number of cars sold, average income), plot the **words on the X-axis** and the **numbers on the Y-axis.** If both sets of data are numbers, plot the **factor being measured** on the *Y*-axis.
- Numbers on the X- and Y-axes are chosen carefully to make the **best use** of the space available. It's not necessary to always begin numbering with zero. It's also not necessary to use the same number scale on both the X- and Y-axes. As shown in **Figure 1-2,** each axis should have a number scale tailored specifically to the data being presented.
- It's not permissible to extend lines or bars **outside the margins** of the graph. Adjust the graph scale to make the data fit comfortably.

3. The data point shown in Figure 1-2 is at the intersection of 10 on the X-axis and 12 on the Y-axis.

4. **Bar graphs** and **line graphs** are examples of types of graphs that are used frequently to present scientific data. **Figure 1-3** illustrates two different ways to present the same information. Note that in both versions of the graphed data:

- Lines or bars are **large and easy to read.**
- Each graph has a **title** that describes the subject matter being graphed. The title can be placed either above or below the graph.
- Each axis has a **title** that clearly explains the information being plotted (including units, if appropriate).
- If the graph contains more than one set of bars or lines (as is the case in **Figure 1-3**), each must be identified with a **key.**

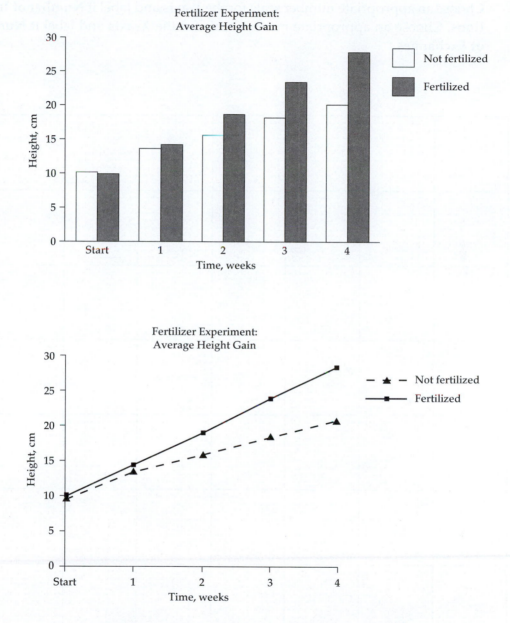

FIGURE 1-3. Comparison of Bar and Line Graphs

✓ Comprehension Check

1. On the graph paper in **Figure 1-4, plot a graph of your experimental results from Activity 4.**

    Choose an appropriate number scale for the **Y-axis** and label it **Number of Infections.** Choose an appropriate number scale for the **X-axis** and label it **Number of Exchanges.**

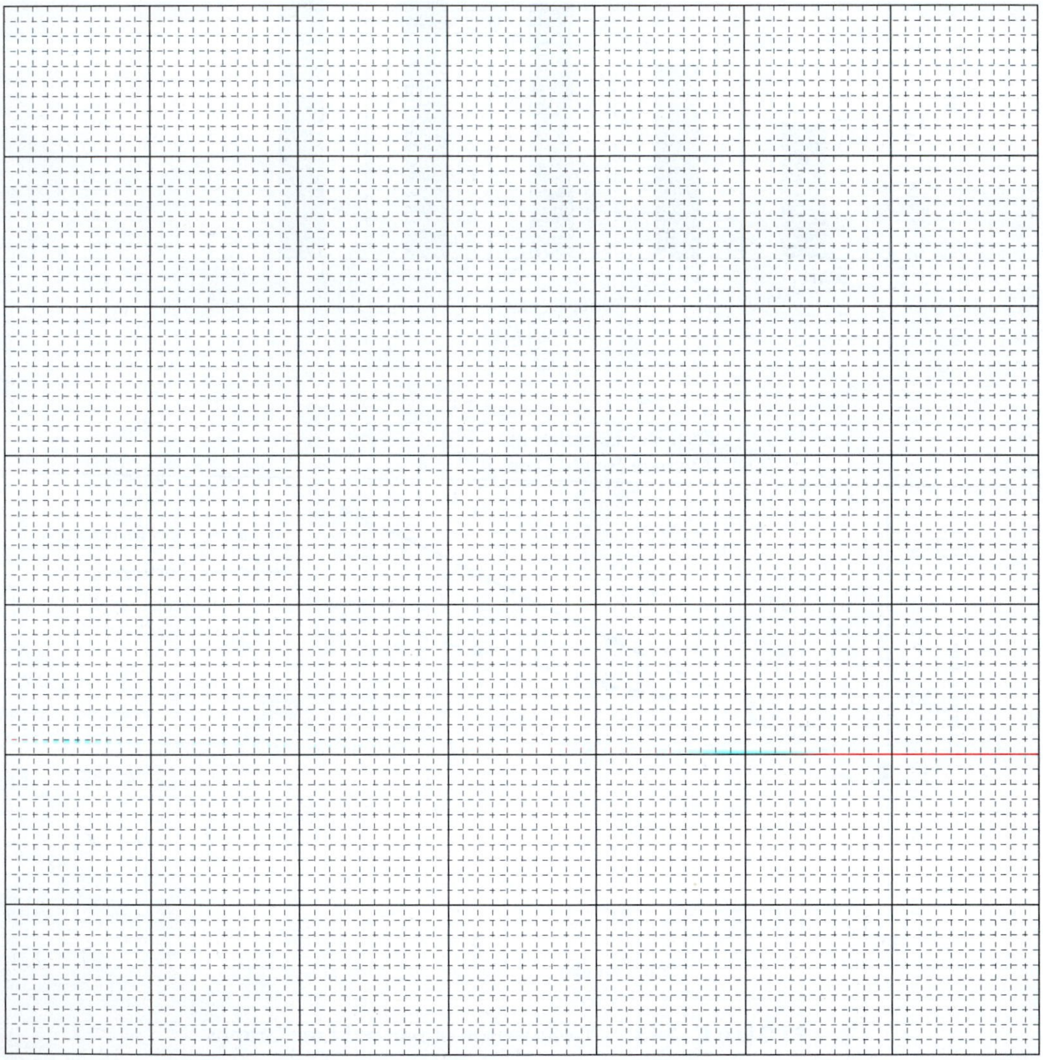

FIGURE 1-4. Results of HIV Simulation

2. In a few sentences, summarize the results of the experiment. In your summary, include data from your summary chart and graph. **Statements of results should include only facts—no interpretation.**

3. In real life, infections don't spread as rapidly as they did in our simulation. Why not?

4. Suggest some methods to slow the rate of infection in the general population.

Check your graph and answers with your instructor before you continue.

ACTIVITY 6

Often, the data collected in an experiment is in a form that isn't easily understandable. Measurements have been established to make it easier to interpret and draw conclusions from large collections of information. We're all familiar with the U.S. Census, which collects huge amounts of information on family size, income, housing conditions, population distributions, and so on.

Simple statistical analysis can reduce the data and convert it to a usable form. A similar approach is used when analyzing the results of large experiments (such as evaluating the effectiveness of new medications or airbags in automobiles). The most commonly used statistical measures are the mean, the mode, the median, the range, and the standard deviation.

The **mean** is the **average** of a set of numbers. The mean is equal to the **sum of all the numbers in the set divided by the sample size.** For example, to find the average pulse rate of a group of 10 students, you would add the pulse values for all of the students together and then divide the answer by the number of students (10).

The mean of a group of numbers often doesn't give you the information you need to correctly interpret the data. **For example, the following two sets of numbers have exactly the same mean, but the spread (dispersion) of numbers is quite different.**

Set 1: 39, 38, 38, 40, 40   mean = 39

Set 2: 3, 29, 25, 38, 100   mean = 39

The **mode** is the **most frequently occurring number** in a set. The mode represents the most common response and, therefore, can be used as a predictor to determine market response (for example, which car model will sell the best in a specific area of the United States).

The **median** is the **middle** number of a set when they're arranged in either **ascending or descending order.** If your income level is above the median, for example, your salary is in the upper 50% of salaries being compared. If a set of numbers has **no middle value,** you can find the median by **averaging the middle two numbers** in the set.

The **range** is the **difference between the largest and smallest values** in a set, for example, the difference between the number of yards gained by the best and worst running backs in the National Football League.

To demonstrate how applying different statistical measures changes the meaning of results, consider the set of 20 biology exam scores in **Table 1-6.**

TABLE 1-6
BIOLOGY EXAM SCORES

STUDENT NUMBER	SCORE: EXAM 1	STUDENT NUMBER	SCORE: EXAM 1
1	90	11	88
2	94	12	54
3	80	13	32
4	82	14	47
5	91	15	25
6	46	16	56
7	97	17	59
8	96	18	60
9	87	19	87
10	84	20	86

Mean = 72.05

Mode = 87

Median = 83

1. What's the **range** of exam scores in **Table 1-6?**

2. If your **exam score was 80,** was your score in the top 50% of the class? **Explain** your answer.

3. In this situation, is the **mean** a good representation of the class scores? **Why or why not?**

4. While watching television last night, you saw an advertisement for Lose-Fast Weight Control Pills. The 12 women shown lost an **average** of 30 pounds while taking the pills. What additional **statistical measures** (as discussed in **Activity 6**) should be reported for you to make an informed decision whether or not to purchase this product?

 Check your answers with your instructor before you continue.

SELF TEST

A national hospital chain performed the study to determine whether the new drug was effective in lowering blood pressure in patients suffering from hypertension (high blood pressure):

Group A (1000 patients) received a daily pill containing the new drug. Group B (1000 patients) received an identical-looking pill, but no medication was included. All the participants in the study had previously been identified as having higher blood pressure than normal (above 140/95 mm Hg). Blood pressure was checked twice daily for a month in all 2000 participants in the study. At the end of the study, the results were published in the *New England Journal of Medicine*.

Fill in the blanks with the choice that is most appropriate. The questions relate to the study of the new blood pressure medication that was described above. **Answers can be used only once.**

a. variable
b. control
c. experimental
d. hypothesis
e. verify
f. X axis

g. Y axis
h. data point
i. mode
j. average
k. median

1. _____ In this experiment, Group A was the _____ group.

2. _____ Group B represents the _____ group.

3. _____ When the results of the study are published, other experts will want to duplicate the experiment in order to _____ the results.

4. _____ If you were graphing the results of this study, on which axis of the graph would it be best to plot blood pressure changes?

5. _____ To calculate the _____ decrease in blood pressure in Group A, you should add up all the numbers in the data set and divide by the number of data points.

6. _____ Each time a patient's blood pressure is measured, it adds one _____ to the study results.

7. _____ The _____ being tested in this study was that the new drug would decrease patient's blood pressure.

Identify the graphing mistakes in **Figures 1-5 and 1-6:**

8.

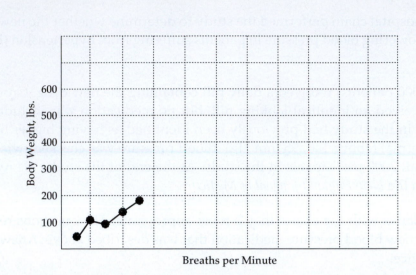

FIGURE 1-5. Sample Graph One

9.

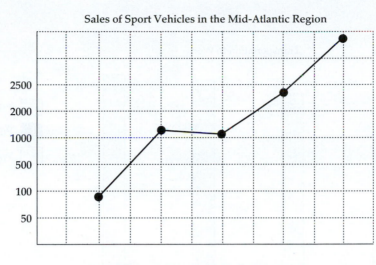

Sales of Sport Vehicles in the Mid-Atlantic Region

FIGURE 1-6.   Sample Graph Two

Answer the following questions in reference to the graph in **Figure 1-7.**

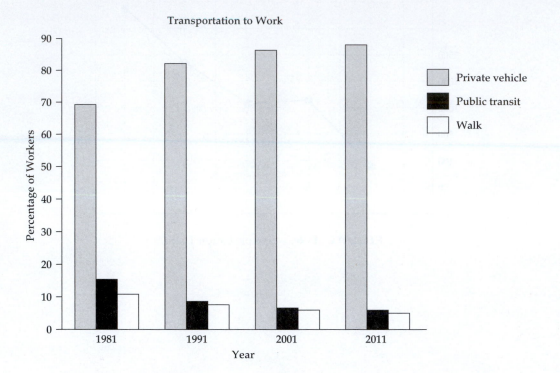

FIGURE 1-7. Percentage of Workers Using Various Modes of Transportation

10.   **(Circle one answer.)** Between 1981 and 2011, the percentage of workers who walked to work **increased / decreased / remained the same.**

11.   Between 1981 and 2011, the number of people who drove cars to work increased by _____ %.

12.   In 2011, what percentage of workers drove cars to work? _____

13.   The number of workers who used public transportation dropped by 7% between 1981 and _____.

14.　　On the graph grid in **Figure 1-8,** graph the following data on energy usage in the home (www.energystar.gov). Graph the data in order from the **lowest percentage** of energy consumed to the **highest.**

HOME ENERGY CONSUMPTION	PERCENTAGE OF TOTAL (%)
hot water heating	14
lighting	12
electronics	4
space cooling	17
space heating	29
appliances	13
other	11

FIGURE 1-8.

Windows to a Microscopic World

Objectives

After completing this exercise, you should be able to:

- identify the parts of the compound and dissecting microscopes and explain their functions
- choose the correct type of microscope for viewing different specimens
- focus the compound microscope using the scanning, low-, and high-power lenses
- prepare a wet mount slide
- correct viewing problems that commonly occur when using the compound microscope
- accurately describe specimens viewed through the dissecting and compound microscopes
- use the microscope to test a hypothesis

CONTENT FOCUS

All organisms, large and small, make valuable contributions to the functioning of the ecosystem. Many plants and animals are large and easy to see, but many important organisms are too small to be seen without the assistance of magnifying lenses.

You'll have an opportunity to get a close look at the anatomy and behavior of some interesting organisms that live around you unnoticed. Microscopes work like magnifying glasses to give you a closer look at these small organisms.

Different types of microscopes can be used for different purposes. During this exercise, you'll be learning to use a **dissecting microscope** to examine larger objects and a **compound microscope** to view smaller specimens. However, other types of specialized microscopes allow us to see even smaller details.

An **electron microscope** focuses a beam of high-energy electrons onto a specimen to create a detailed image. Depending on the type of electron microscope, you can examine objects on a very fine scale, viewing both the internal structures and details of external appearance (see the example in **Figure 2-1**).

FIGURE 2-1.   A Scanning Electron Micrograph of a Bedbug

ACTIVITY 1 — LEARNING TO USE THE DISSECTING MICROSCOPE

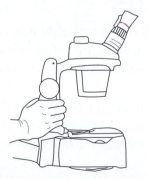

Caution!
Always carry a microscope upright with one hand under the base and the other hand on the arm.

1. Work in groups of **two students.** Get the following supplies: **a dissecting micro-scope and a penny.**

2. Place the penny on the microscope **stage.** Locate the stage by referring to the photo in **Figure 2-2.** Turn the penny so that the Lincoln Memorial is facing you. **Adjust the light** until the penny is brightly illuminated.

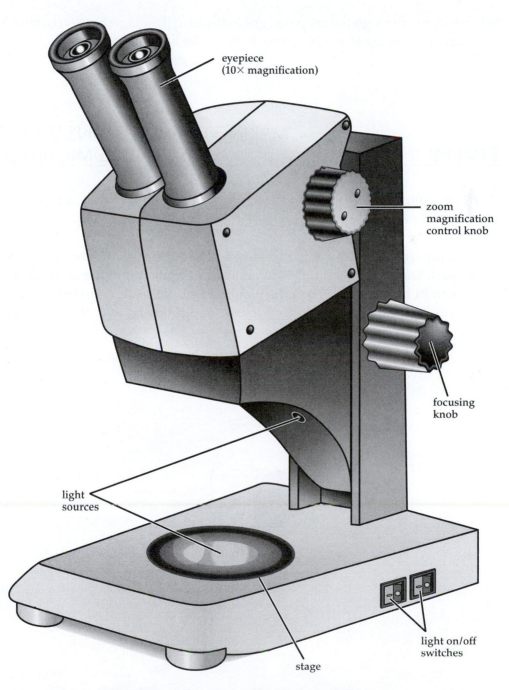

eyepiece
(10× magnification)

zoom
magnification
control knob

focusing
knob

light
sources

light on/off
switches

stage

FIGURE 2-2.   Parts of the Dissecting Microscope

3. Set the **magnification control knob,** located on the top or the side of the **head,** to the **lowest setting.**

4. Turn the **focusing knob,** located on the microscope **arm,** until the head is as close to the stage as possible. Look through the **ocular lenses (eyepiece)** at the penny. **Turn the focusing knob** until the image of the penny is sharp and clear (the head will be moving **away** from the stage).

   Is Lincoln sitting in the Lincoln Memorial in your penny? _____

5. While looking through the eyepiece, gradually turn the magnification control knob. The change in image size will resemble the zoom action of a camera.

ACTIVITY 2
MAKING OBSERVATIONS WITH THE DISSECTING MICROSCOPE

1. Work in groups. Get the following supplies: **a small glass bowl, a pipette,** and **a blunt metal probe.**

2. Use the pipette to remove a **planaria** from the culture container and place it in your bowl. Locate the worm of your choice and draw it into the tip of the pipette (as shown in **Figure 2-3**). When the planaria feels the water current, it will probably form itself into a small protective ball, and it will be easy to draw it into the pipette.

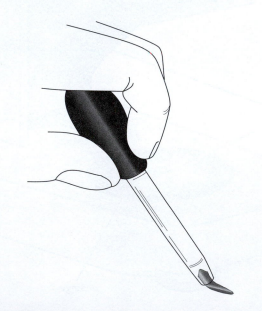

FIGURE 2-3. Drawing a Planaria into the Pipette

Caution!
Don't suck the planaria too far into the pipette. It will attach to the inside of the pipette and you won't be able to get it out.

If additional water is needed in your glass dish, add pond water. Don't use tap water or distilled water. |

3. **Observe your worm** under the **dissecting microscope.** Use a medium or low light level.

   Observe carefully through the microscope for at least **two minutes.**

   Does the worm swim or crawl? _____

   Describe the movement of the worm's **muscles** as it moves forward.

4. **Draw a picture** of your worm. **Increase the magnification** in order to see the details of the head. When drawing the head, pay particular attention to the position of the eyes in relation to each other. Make your drawing **large** and **clear. Label** the **head** and the **eyes.**

DRAWING OF PLANARIA

5. Hold the **blunt probe motionless** directly in the path of a moving worm. How does the worm respond?

6. Touch the head end **gently** with the blunt probe. How does the worm respond?

7. Touch the planaria **gently** on several other body parts. Record your observations.

8. In what way was the **touch response** on other body parts **similar** to or **different** from the response you observed when the worm was touched on the head end?

 Check your answers with your instructor before you continue.

ACTIVITY 3

GETTING FAMILIAR WITH THE COMPOUND MICROSCOPE

1. Work in groups. Get a compound microscope.

Caution!
Always carry the microscope upright with one hand under the base and the other hand on the arm.

2. Plug the microscope into an electrical outlet and **turn the light on.** The light switch is located on the **base** of the microscope.

3. The compound microscope consists of a system of **optics** (lenses and mirrors) and focusing controls. The base and the arm support a **body tube** that **houses the lenses** that magnify the image.

4. At the top of the scope is the **ocular lens (eyepiece).** The ocular lens is only one of a series of lenses that magnify the image. The ocular lens makes the image **10 times larger** than life size (abbreviated **10×**).

    Look through the ocular lens. You'll see a black pointer. Rotate the eyepiece **while looking through the lens.**

    What happens to the pointer? _____

5. At the bottom of the body tube is a revolving **nosepiece.** The lenses that screw into the nosepiece are called **objective lenses.**

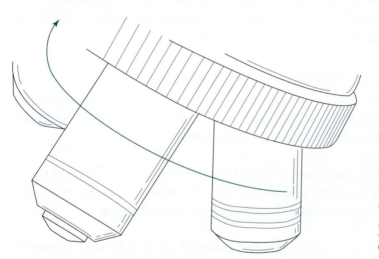

Turn the nosepiece until you hear or feel one of the objective lenses quietly **click into position** (see **Figure 2-4**). When an objective lens clicks into position, it's in the proper alignment for light to pass from the light source, through the objective lens, through the ocular lens, and into the viewer's eye. Turn the nosepiece again to bring a different objective lens into position.

FIGURE 2-4.   Revolving Nosepiece and Objective Lenses

6. **How will you know** when you've placed the objective lens in the proper position?

7. Note that each objective lens is of a different length. Each of the lenses has a different magnifying power. The **shortest lens** has the **lowest magnifying power** and the **longest lens** has the **highest magnifying power.**

Because light from the specimen passes through **both** an objective lens **and** the ocular lens, the **total magnification** of the image is the result of the objective lens magnification **multiplied** by the ocular lens magnification.

8. Using the information provided in step #4, **complete the magnification table.**

Length of Objective Lens	Magnification of Objective Lens	Magnification of Ocular Lens (Eyepiece)	Total Magnification of Specimen
Short	4×		
Medium	10×		
Long	40×		

9. On both sides of the arm, you'll see the **focus knobs.** The larger ring is the **coarse focus knob** and the smaller ring is the **fine focus knob.**

   The coarse focus moves the stage in fairly large increments to bring the object on the slide into approximate focus. The small adjustments possible with the fine focus knob are then used to make the image sharp and clear.

10. The **mechanical stage** is a movable platform designed to hold a microscope slide. Notice that one side of the metal stage clip makes a 90° angle. When you place a slide on the stage, **make sure** that one corner of the slide **fits exactly into that angle.** If not, you won't be able to move the slide or focus properly.

    Two control knobs move the mechanical stage. One moves the stage left and right; the other moves the stage forward and back. **Always move the stage using the control knobs. Don't attempt to move the slide using your fingers.**

11. In the center of the stage, you'll see the glass of the **condenser lens,** which focuses the light on the specimen.

12. **Looking at the condenser lens,** move your hand to the **iris diaphragm lever** (just under the stage). Push the lever all the way to the left and then all the way to the right. What happens to the amount of light passing through the condenser lens as you move the iris diaphragm lever?

ACTIVITY 4   PARTS OF THE COMPOUND MICROSCOPE

Now that you have some experience with the parts of the microscope and their functions, **label the parts** of the microscope on **Figure 2-5.**

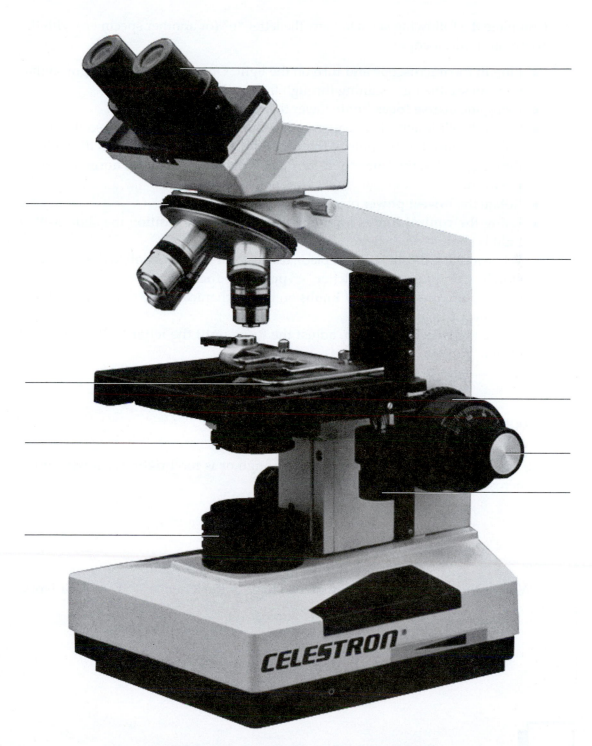

FIGURE 2-5.   Parts of the Compound Microscope

ACTIVITY 5

LEARNING TO USE THE COMPOUND MICROSCOPE

1. Work in groups. Get the following supplies: **a prepared slide labeled letter "e"** and **a piece of lens paper.**

2. Complete the following steps to view the letter "e" (or another specimen) with the compound microscope:

    ■ Plug in the microscope and turn on the light. Open the iris diaphragm so that you can see the light shining through the condenser lens.

    ■ Using the **coarse focus knob,** lower the stage as low as it can go.

    ■ Clean the slide with lens paper and carefully position the slide so that it fits precisely into the 90° angle of the metal clip on the stage. Position the slide on the stage so that the letter "e" is facing you **right-side up,** in its **normal reading position.**

    ■ Rotate the **lowest power** objective lens into position.

    ■ Using the **control knobs on the mechanical stage,** position the slide so that light is shining on the letter "e."

    ■ While looking through the eyepiece, use the **coarse focus knob** to raise the stage slowly. Continue until the letter "e" **pops into focus.**

    ■ Once again, use the **control knobs** on the mechanical stage to **center** the letter "e" in your field of view.

    ■ Using the **fine focus knob,** adjust the focus until the letter "e" is sharp and clear.

✔ Comprehension Check

1. If the letter "e" isn't lighted brightly enough or is too bright, what part of the microscope would it be best to adjust?

2. What is the **total magnification** of the image of the letter "e"? **Show** your work.

Check your answers with your instructor before you continue.

3.   Draw the letter "e" **exactly** as it appears under low power. Make it **large** and **clear**.

LETTER "e"

4.   How is the **orientation** of the letter "e" as seen through the microscope **different** from the way an "e" **normally** appears? List **two** differences.

5.   **While looking through the eyepiece,** move the stage to the **left.**

In what direction does the image appear to move? _____

While looking through the eyepiece, move the stage **away from you.**

In what direction does the image appear to move? _____

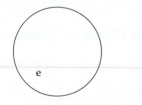

If you wanted to **center** the letter "e" in this drawing, in which **two directions** would you have to move the stage?

_____ and _____

6.   Follow these steps to view a specimen at a higher magnification:

   ■ Center the letter "e" (or other specimen) in your field of view.
   ■ Rotate the **10×** objective lens into position. **DON'T adjust the stage at this point!**
   ■ **DON'T use the coarse focus knob!** Using the **fine focus knob,** adjust the focus **slowly** until the letter "e" is sharp and clear.
   ■ Repeat the previous steps to change to other objective lenses.

7. **Centering the specimen is absolutely necessary before you can change to a higher-power objective!** Do you know why? The reason is simple. The more powerful the magnifying lens, the smaller the area you see, but you see that small area in greater detail. The area you can see at one time is called the **field of view.** So you could say that the **higher** the power of the objective, the **smaller** the field of view.

   If the specimen you're trying to view isn't centered before you switch to a higher-power lens, it may no longer be within the field of view, and you'll think your specimen has disappeared!

8. Looking through the eyepiece, can you still see the letter "e"? _____

   Is the letter "e" exactly in the center of the field of view? _____

   If not, **move the slide slightly** to center the image.

   Is the letter "e" sharp and clear? _____

   If not, **gradually adjust the fine focus knob** until the problem is corrected.

   Do you have enough light? _____

   If not, **gradually adjust the iris diaphragm.**

   How has the **image of the letter "e" changed** from the way it looked using the **4✕** objective lens?

9. Repeat the instructions in step #6, this time changing from **10✕** to the **high-power objective.**

   Looking through the eyepiece, can you see the entire letter "e"? _____

10. **(Circle one answer.)** Your field of view on high power is **larger / smaller** than the field of view on 10✕.

    How does the size of the field of view determine how much of the letter "e" you can see?

11. When using the **high-power objective,** what is the **total magnification** of the image of the letter "e"? _____ **Show your work.**

Check your answers with your instructor before you continue.

ACTIVITY 6
PREPARING TEMPORARY SLIDES— WET MOUNTS

A wet mount is a method of preparing a slide that will be used only for a short time. Unlike the letter "e," which was permanently attached to the slide, a wet mount is made by placing the specimen into a drop of liquid on a slide. The specimen and water droplet are held in place by a coverslip.

Human Epithelial Cells (Cells from the Inside of Your Cheek)

1. Work in groups. Your group will make **two different wet mounts** of cheek cells.

    One will be made with a drop of **stain.** The second will be made by substituting a drop of **physiological saline** for the stain.

2. Get the following supplies: **a dropper bottle of iodine stain or physiological saline, a slide, a coverslip, lens paper,** and **a clean toothpick.**

3. **Place a drop of iodine stain or a drop of saline near the center of a clean slide.**

4. **Gently** scrape the inside of your cheek with the **end of a toothpick.**

Caution!
If you scrape too hard, you'll be examining blood cells instead of epithelial cells!

You've removed some of the cells that form a **protective covering** for the inside of the mouth. Like other epithelial cells, these are constantly being worn off and replaced by new cells of the same type.

5. **Spread the material from the toothpick** into the drop of iodine stain or saline. Add a coverslip.

    Put two microscopes together (yours and your partner's) on the laboratory table so that you can view and compare the stained and unstained slides.

Hint:

Unstained cells are clear. They're only visible with very low light levels. If you think there are no cells on your slide, adjust the iris diaphragm.

---

6. **View both slides. Begin with the 4× objective.** Continue until you've located the cells using **all three objective lenses.**

7. **Draw a picture** of **one** cheek cell, **viewed on high power.** Make it **large and clear.**

HUMAN CHEEK CELL

8. **Label** the following cell structures (referred to as **organelles**) in your drawing of the cheek cells:

Organelles to Label	Function
nucleus	directs all cell activities
cell membrane	controls movement of materials in and out of the cell
cytoplasm	jelly-like fluid found between the nucleus and cell membrane

9. Give an example of a **medical procedure** in which epithelial cells are scraped from another area of the body and **examined microscopically.**

10. Is there an advantage in **using a stain** to view cells microscopically? _____ If so, how is the stain helpful?

Daphnia—An Aquatic Organism

The water flea, daphnia, is a microscopic organism commonly found in ponds, lakes, and streams. Because they are small and transparent, living daphnia can be studied easily in the laboratory. Daphnia feed on microscopic food particles. Their five pairs of legs are modified into strainers that filter the food particles from the water. Daphnia, in turn, are an important part of the food chain. Many fishes and even larger aquatic animals feed on daphnia.

1. Work in groups. Get the following supplies: **a depression slide** and **a bottle of methyl cellulose** (the bottle may also be labeled **Protoslo®**).

2. Put a **small** drop of the methyl cellulose into the depression on your slide.

    Use the pipette in the culture jar to remove a daphnia from the container and place it in the depression on the slide. **You don't need a coverslip.**

3. Make your observations using the **scanning lens** of the microscope **(4×)**. Add the following details to the daphnia outline in **Figure 2-6: eye, heart, brain, and intestine.**

Label the details you're adding to the drawing. Also, label the head and legs.

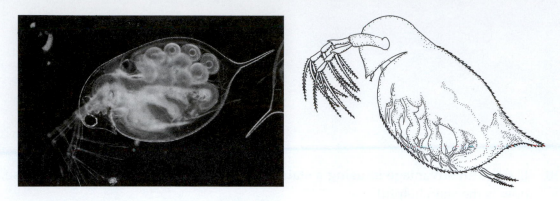

FIGURE 2-6.   Daphnia

Check your answers with your instructor before you continue.

ACTIVITY 7
USING THE MICROSCOPE TO ANSWER A QUESTION: WHAT DO AQUARIUM SNAILS EAT?

1. Work in groups. Go to your **classroom aquarium** and **observe the feeding activity** of the snails.

2. **Write a hypothesis:** What are the snails **eating?**

   Hypothesis:

Hint:
Don't forget to make sure your hypothesis is testable!

3. Develop and carry out an observation plan. Refer to **Figure 2-7** to help you identify organisms you observe. **Record** the results of your observations.

✔ Comprehension Check

1. **Discuss** the results with your other group members. **(Circle one answer.)** My hypothesis **was / was not** supported.

2. Write a conclusion based on your hypothesis and collected data. Support your conclusion by mentioning facts collected during your experiment.

Check your answers with your instructor before you continue.

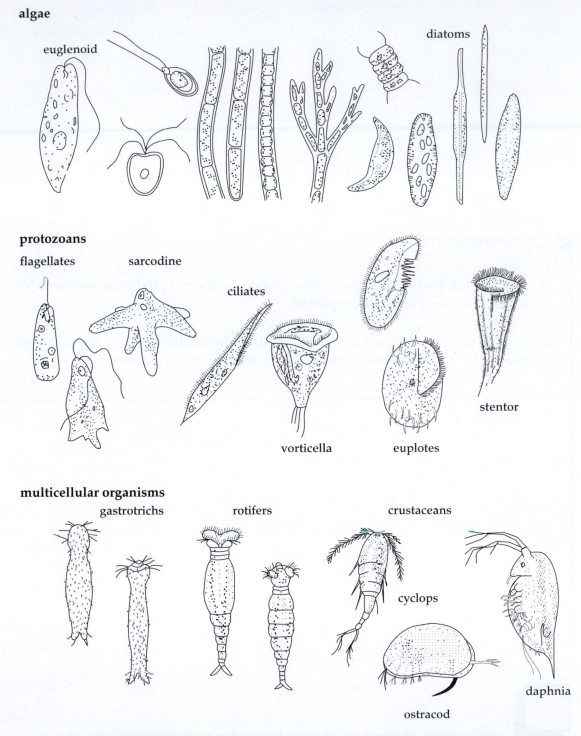

algae

euglenoid

diatoms

protozoans

flagellates

sarcodine

ciliates

stentor

vorticella

euplotes

multicellular organisms

gastrotrichs

rotifers

crustaceans

cyclops

ostracod

daphnia

FIGURE 2-7. Organisms Commonly Found in Pond Water

SELF TEST

1. Complete the table by entering the appropriate part of the compound microscope or the correct function for the part of the microscope listed.

Parts of the Compound Microscope	Function
Ocular lens (eyepiece)	
Stage	
	Focuses light on the specimen
Nosepiece	
	Objective lens used to first locate a specimen
	Regulates the amount of light that passes through the specimen
Fine focus knob	
	Objective lens with the lowest magnifying power
	Objective lens with the highest magnifying power
Coarse focus knob	

2. What is the **total magnification** if the ocular lens is 15× and the objective lens is 20×? **Show your work.**

3. List **three differences** between the dissecting microscope and the compound microscope:

    a.

    b.

    c.

4. **Which type of microscope** (compound or dissecting) would you use to observe the following:

    _____ Cells from the lining of your stomach

    _____ A seashell you found on the beach

    _____ A cockroach you found in the kitchen

    _____ Mold from your shower curtain

5. Suppose you were watching a daphnia under the compound microscope and noticed that it moved **toward you** and then **to your right.**

    Which direction(s) did the daphnia **actually** move? _____

    Explain your answer.

6. Explain how you would **correct the following problems** experienced when using a microscope:

   a. When changing magnification from **10✕** to **40✕**, the specimen disappears.

   b. The field of view is too dark.

   c. Your field of view is partially obscured by a dark area.

   d. There's a fingerprint in your field of view.

   e. There are many hollow, dark circles in your field of view.

Functions and Properties of Cells

Objectives

After completing this exercise, you should be able to:

- identify and explain the functions of the major cellular organelles
- explain the similarities and differences between prokaryotic and eukaryotic cells
- explain the similarities and differences between plant and animal cells
- apply your knowledge of cell structure and function to real-life situations

CONTENT FOCUS

Typically, when you look at a plant or animal, it's easy to think of an organism as one large unit. When looking through the microscope, however, it becomes obvious that an organism is composed of trillions of tiny units called **cells,** which work cooperatively to carry out the functions that keep us alive. If the activity of your cells stopped, even for a few moments, death could quickly follow.

Within each cell lies a collection of specialized structures called **organelles,** in which particular chemical activities take place. The various cell structures carry out activities that mirror the functions of our body organs.

Organisms can be designated as either prokaryotic or eukaryotic depending on the structure of their cells. The most familiar example of organisms with **prokaryotic cells** is bacteria. Plants and animals are composed of **eukaryotic cells.** About 200 different types of eukaryotic cells are found in the human body. Prokaryotic and eukaryotic cells can be quite different in shape, structure, and function, but they all have some basic characteristics in common.

ACTIVITY 1   CELLULAR ORGANELLES

Learning about cell organelles will be more meaningful when you understand the role that each part plays in the life of a cell. Often the presence of organelles is **related to the function** of various cells. In a cell with a specific function, **some organelles may be more numerous** than in other nearby cells with different functions.

Some cell organelles are large enough to be seen with your compound microscope. Others are much smaller, but they can be seen and photographed with more powerful microscopes. Also, although **most organelles are found in both plant and animal cells,** there are some **exceptions.**

Table 3-1 gives a brief summary of the functions of some basic organelles found in eukaryotic cells. Based on the information in the table, **fill in the blanks with the appropriate organelle or organelles.** Answers can be used **more than once.**

1. Three organelles that are found in plant cells, but not in animal cells: _____, _____, and _____

2. Organelle where muscle proteins are manufactured. _____

3. Provides strength and support in tree trunks. _____

4. An infertile man has a sperm sample tested and discovers the sperm have low motility (don't swim normally). Name the malfunctioning organelle _____

5. Two structures that a material would have to cross to enter the cytoplasm of a plant cell. _____ and _____

6. Would be more numerous in muscle cells than in skin cells. _____

7. Would be functioning when fat is manufactured for storage in a cell. _____

8. Location where genetic information is stored. _____

9. Location in the cell where most organelles can be found. _____

10. Allows some materials to enter the cell, but not others. _____

T A B L E   3 - 1

SOME CELL ORGANELLES AND THEIR FUNCTIONS

Organelles Visible with a Light Microscope	Function
Cell wall	External support and protection; if present, located outside the cell membrane; composed primarily of cellulose; not found in animal cells
Cell membrane	Surrounds the cytoplasm; barrier between the environment and the cytoplasm; controls the movement of materials in and out of the cell
Cytoplasm	Liquid/gel "filler" substance inside the cell membrane; all internal cell organelles are suspended in the cytoplasm
Nucleus	Stores and transfers information (DNA and RNA) needed to control cell functions
Chloroplast	Structure in which photosynthesis takes place; contains chlorophyll and is green in color; not found in most animal cells
Central vacuole	Membrane-enclosed bag of fluid; water storage organelle in a mature plant cell; not found in animal cells
Cilia or flagella	Hair-like projections from the cell membrane; capable of movement; assist in cell locomotion or move materials across the cell surface
Organelles Not Visible with a Light Microscope	**Function**
Mitochondria	Location of aerobic cellular respiration; produce ATP energy, which is used by cells to do work
Ribosomes	Protein synthesis; ribosomes may be found free in the cytoplasm or attached to the endoplasmic reticulum (ER)
Endoplasmic reticulum (ER)	Network of membranes important for transport of molecules within the cytoplasm and across the cell membrane; many chemical reactions occur here, including synthesis of some carbohydrates, lipids, and proteins; divided into "rough" and "smooth" sections with different functions
Smooth ER	Carbohydrate and lipid synthesis
Rough ER	Protein synthesis; ribosomes are attached to the ER membrane in this location
Golgi apparatus	Packaging of molecules for export out of the cell; most exported molecules are proteins

Prokaryotic cells have some similarities and some differences from eukaryotic cells. Both types of cells are enclosed by a cell membrane. Most prokaryotic cells also have a rigid cell wall that surrounds and protects the cell.

Prokaryotic cells don't have a nucleus, but a region in the cytoplasm called the **nucleoid region** is where one or more chromosomes are located. Because the nucleoid region isn't enclosed by a membrane, it isn't considered to be a true nucleus. The term "prokaryotic" means "before the nucleus" and is a reference to the fact that prokaryotic cells evolved earlier in earth's history than the more familiar eukaryotic cells found in our bodies. In some prokaryotes, DNA can also be found outside the nucleoid region in small structures called **plasmids.** Plasmid genes can be transferred between cells, passing traits such as antibiotic resistance from one cell to another.

Other types of membrane-enclosed organelles are also absent in prokaryotic cells. The functions performed by eukaryotic organelles occur in the cytoplasm of prokaryotic cells or along sections of the cell membrane that are folded into the cytoplasm.

11. Based on the diagram in **Figure 3-1,** which cellular organelles found in eukaryotic cells are also present in prokaryotic cells?

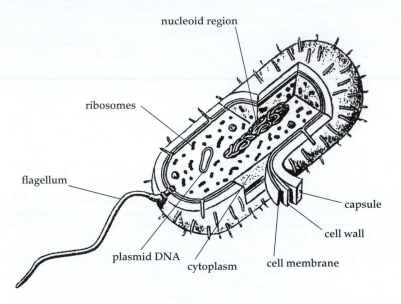

FIGURE 3-1.   Structure of a Prokaryotic Cell

12. What do prokaryotic cells have instead of a nucleus?

Fill in the blank with the most appropriate cell type. Questions may have more than one answer. Answers can be used more than once.

a. plant cell

b. animal cell

c. bacterial cell

13. _____ You observed some cells through a microscope. You saw ribosomes but no mitochondria.

14. _____ You saw a cell with a flagellum, but it wasn't a sperm.

15. _____ You saw a cell with a cell wall. Which choice would be an incorrect definition?

Check your answers with your instructor before you continue.

ACTIVITY 2   OBSERVING LIVING CELLS

1. Work in groups. Get the following supplies: **slides, coverslips, pipettes, forceps, and lens paper.**

2. **Amoeba** is a **one-celled organism** commonly found in pond water.

    Elodea is a **freshwater plant** frequently used in home aquariums. You'll be looking at an elodea leaf, which is thin enough to use with the compound microscope. An elodea leaf has two layers of cells. Each cell looks like a small rectangle.

    Get **two compound microscopes** and set them up side by side. Following directions from your instructor, make **two wet mounts,** one of an **amoeba** and one of an **elodea leaf.** Place **one wet mount on each microscope.**

3. **Figure 3-2** contains a photograph of an amoeba. Based on your microscope observations and the photo in the figure, **label** all the organelles you can identify on the **photo of the amoeba.**

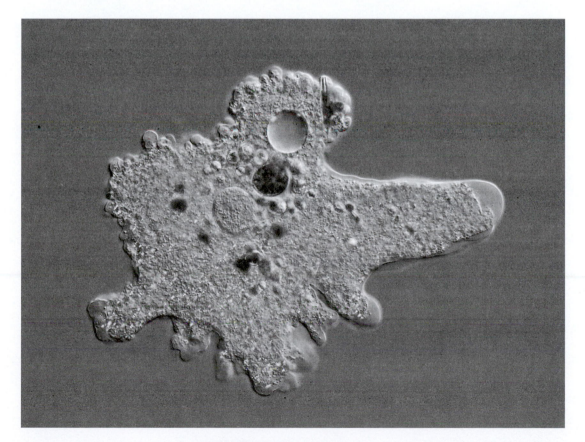

FIGURE 3-2. Amoeba

4. **Figure 3-3** contains photographs and a diagram of elodea cells. Based on your microscope observations and the photos in the figures, **label** all the organelles you can identify on **both the diagram and photo of the elodea**.

Hint:

Use the information in Table 3-1 to help you locate and identify the cellular organelles.

FIGURE 3-3A. Elodea Leaf

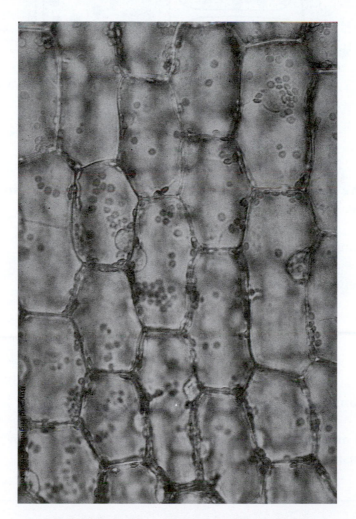

FIGURE 3-3B. Photograph of Elodea Cells

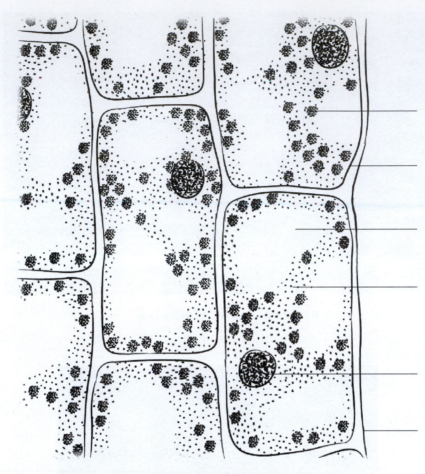

FIGURE 3-3C.   Diagram of Elodea Cells

Answer the following questions based on your observations through the microscope:

5.   Which organism(s) can carry on photosynthesis? _____

What **evidence** did you see that led you to this **conclusion**?

6.   Which organism has a **cell wall?** _____

7.   Which **cell organelles** can be seen changing position in the moving cytoplasm of each organism?

Elodea _____

Amoeba _____

8. Some organisms can secure food by surrounding their prey with cell extensions. This process is called **phagocytosis** (shown in **Figure 3-4**).

   Which organism(s) do you think would be able to do this? _____

   Explain your answer.

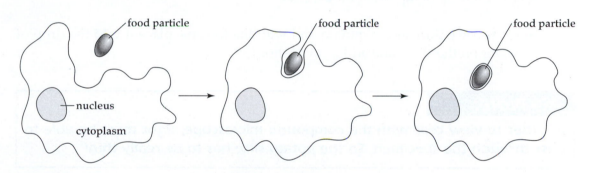

food particle food particle food particle

nucleus

cytoplasm

FIGURE 3-4. Phagocytosis

9. Which organism demonstrates **locomotion** (ability to move from place to place)?

   _____

Check your answers with your instructor before you continue.

ACTIVITY 3   AMYLOPLASTS IN POTATO TUBERS

Chloroplasts in leaf cells (such as the elodea leaf) carry out the process of photosynthesis and form sugar (glucose). Excess glucose can be converted into a stored form (starch) not only within the leaves, but also in other locations in the plant.

   Amyloplasts are cellular organelles that synthesize and store starch in various locations in the plant. The most common locations to find amyloplasts are in fruits, roots, and underground stems modified for energy storage, such as potatoes. When the plant needs energy, the amyloplasts convert the stored starch back into glucose.

1.   Work in groups. Get the following supplies: **a compound microscope, a fresh piece of potato, a scalpel or razor blade, forceps, a dropper bottle of iodine, a couple of pieces of lens paper, slides, and cover slips.**

2.   To prepare your potato wet mount:

   ■ Set the slide on a paper towel (so you can see what you're doing).

   ■ Place a drop of tap water on the slide.

   ■ Cut a **very thin** slice of potato and with the forceps, place it into the drop of water on the slide, and add a cover slip.

> ### Note:
> **In order to view cells with the compound microscope, light must be able to pass through the specimen. So the potato slice has to be really thin!**

3.   Place the slide on the microscope. Start with the **4×** scanning lens and dim the light level with the iris diaphragm until the cells can be clearly seen. Center and focus; then change to low power (**10×**). Readjust the light level and focus as necessary.

   The multiple amyloplasts within the potato cells will look like clear ovals. In addition to the amyloplasts, what **other cellular organelles** can you identify in the potato cells?

4.   Iodine is an indicator that shows the presence of starch by turning black. Add iodine to your potato slice to view the amyloplasts as follows:

   ■ Leave the slide in position on the microscope and **don't lift the cover slip** or adjust the lenses.

   ■ Add a **drop of iodine** to one edge of the cover slip.

   ■ Take a small piece of lens paper and place it with one edge under the opposite
     side of the cover slip from the iodine drop.

   ■ The lens paper will begin to absorb the water from your wet mount, automati-
     cally drawing the iodine onto the potato slice.

   ■ If needed, add more iodine to the edge of the cover slip.

5. What was the color of the iodine you added to the edge of your cover slip?
    _____

    What color did the iodine become when it came in contact with the potato cells?
    _____

    What does this color change indicate? _____

    Within which cellular organelle did the color change occur? _____

6. Based on your observations, draw a large picture of a few representative potato
    cells in **Figure 3-5.** On your drawing, label the following structures: **cell wall, cyto-
    plasm, amyloplasts,** and **starch granules.**

POTATO CELLS

FIGURE 3-5.   Potato Cells

7. **Challenge Question!** Amyloplasts and chloroplasts have similar cellular origins and are closely related in structure. As a result, amyloplasts can turn into chloroplasts under certain environmental conditions. What do you think might happen if potatoes were exposed to sunlight for a few days? Why?

ACTIVITY 4   STEM CELLS

You have seen that cells of the body are specialized for their particular functions. How can so many different types of body cells arise from a single cell—the fertilized egg? The first cells that develop in an embryo are called **embryonic stem cells.** These cells are unspecialized and have the ability to develop into any type of body cell.

The fertilized egg contains your entire DNA library (your genes). Genes can be turned on or off within a cell, causing that cell to develop differently and specialize for different functions in the body. The process is called **gene regulation.**

Embryonic stem cells can develop into any type of body cell, but most cells in adults have lost this ability. The stem cells that do remain in adults can produce only a few different types of cells for the tissues in which they are located. Adult stem cells are found in the brain, bone marrow, body muscles, and skin.

1. Susan has been diagnosed with leukemia, a form of cancer. Treatment often requires the destruction of abnormal cells in the bone marrow. In the process, normal cells are usually destroyed as well. Susan has been offered a bone marrow transplant from a healthy donor. How will this help her condition? In your explanation, use the following terms: **blood cells, adult stem cells, cell division,** and **gene regulation**.

SELF TEST

Fill in the blanks in **questions 1–9** with the correct **organelle name** or **organelle function.**

1. You went on vacation and forgot to water your plants. When you returned, the leaves were soft and wilted. Name the organelle that became dehydrated. _____

2. If I noticed that some skin cells contain an extensive network of smooth ER, I could hypothesize that the cell was synthesizing large amounts of _____.

3. If a scientist wanted to remove genetic material from a cell to make a DNA fingerprint, what organelle could be used? _____

4. If a cell lacked ribosomes, what type of molecule synthesis would be absent? _____

5. A cell without a cell wall is probably also missing which organelle(s)? _____

6. List all the organelles that are involved in the production, packaging, and export of proteins from a cell to other parts of the body.

7. In lab, you decided to grind up a leaf to see which cell organelles you could identify. Some organelles on the filter paper seemed to be absorbing carbon dioxide and releasing oxygen. The organelles are most likely _____.

8. The energy that sperm use to swim is produced by _____.

9. Paramecium, a common one-celled pond organism, is covered with tiny little "hairs" that it uses to swim. The scientific name for these hairs is _____.

10. Of the cell organelles studied in this exercise, **list three** that are found in plant cells but not in animal cells.

11. Based on your knowledge of photosynthesis, why are central vacuoles important organelles in plants?

12. In an accident in a chemical plant, cyanide gas was accidentally released. It's known that exposure to cyanide affects production of the energy-rich molecule ATP. Which cellular **organelle** would be affected by the poison? **Explain** your answer.

Movement of Molecules Across Cell Membranes

Objectives

After completing this exercise, you should be able to:

- explain the concepts of diffusion and osmosis and why they are important to cell physiology
- use indicator chemicals to test for the movement of molecules and to determine the direction of diffusion
- explain the process of osmosis in living cells exposed to different extracellular solute concentrations

CONTENT FOCUS

There are several ways that nutrients, ions, oxygen, carbon dioxide, and other molecules can be moved in and out of cells across the cell membrane. Two of these transport methods, **diffusion** and **osmosis,** take advantage of the fact that molecules are in constant motion.

The process of **diffusion** occurs whenever dissolved particles move from an area of **high concentration** (more of them) to nearby areas where they are **less concentrated.**

Cells are surrounded by a membrane that allows some substances to pass through, but not others. This is referred to as a **selectively permeable membrane.**

A process similar to diffusion occurs with **water** molecules in the environment. Water molecules disperse through a **selectively permeable membrane** from an area of **high concentration** to an area of **lower concentration.** This process is given a different name, **osmosis.**

ACTIVITY 1   WATER MOLECULES IN MOTION

When you looked at cells through the microscope, you observed the movement of cytoplasm. Everywhere around us, molecules of liquids and gases are in constant, random motion.

Of course, these moving molecules are much too small to be observed directly, but we can demonstrate this activity by placing some colored powder into a drop of water.

1. Work in groups. Get the following supplies: **a slide, a cover slip, a dropper bottle of distilled water, a toothpick, and carmine powder.**

2. Place a drop of distilled water on the slide. **Carefully** add a **tiny** amount of carmine powder to the water drop. Use the **pointed tip** of the toothpick to scoop up the carmine powder.

> ### Hint:
> **Imagine how much carmine powder you think you might need, and then scoop up half that amount.**

Add a cover slip and observe the slide under **low power,** then **medium power,** and finally, **high power.**

3. **Describe** the activity of the carmine particles in the water droplet.

4. Considering that carmine particles are **not** alive, what's **causing** the carmine particles to move?

As was mentioned earlier, molecules of liquid are always in motion. This means water molecules everywhere are constantly moving and bumping into each other.

As they move, they also collide with anything floating in the water, such as the carmine particles.

(Circle one answer.) If we warmed up the carmine slide a little bit, the rate of molecular motion would be **faster / slower / stay the same. Explain** your answer.

The process of **diffusion** occurs whenever dissolved particles move from an area of **high concentration** (more of them) to nearby areas where they are **less concentrated.** The diffusion process is similar to a bumper car ride (see **Figure 4-1**). Cars are moving quickly and often collide. The force of the collision sends the cars shooting outward into an empty area of the bumper car rink. The more an area is crowded with bumper cars, the more collisions occur.

In this way, the cars are gradually dispersed from the crowded center of the rink (an area of **high** bumper car concentration) to the emptier fringe areas (areas of **lower** bumper car concentration). This is a good model of what happens when particles diffuse in the environment (see **Figure 4-1**).

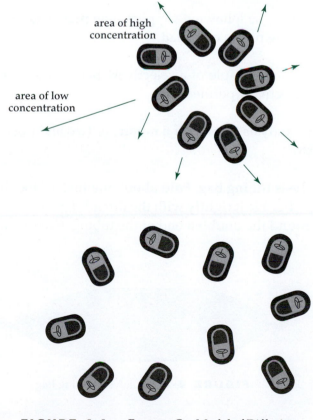

FIGURE 4-1. Bumper Car Model of Diffusion

Diffusing molecules always move outward from an area of high concentration into areas of lower concentration. The difference in concentration levels between two locations (for example, inside the cell compared to outside the cell) is known as the **concentration gradient.** When molecules move from an area of high concentration to an area of lower concentration, we say they are moving **with the concentration gradient.**

Water molecules disperse through a **selectively permeable membrane** from an area of **high concentration** to an area of **lower concentration.** Although the mechanism is similar to that of diffusion, the process is given a different name, **osmosis.**

Solutions used by cells have two components: dissolved molecules (called **solutes**) and water (the **solvent**). The cell membrane, which controls the movement of materials in and out of the cell, can't control the movement of all molecules. **Carbon dioxide, oxygen,** and **water** are examples of molecules that move passively through the cell membrane following the concentration gradient.

We can demonstrate the processes of diffusion and osmosis in an experiment using **dialysis tubing,** an artificial membrane with many small pores. We'll be testing a solution with **three solutes: sugar, salt, and starch.**

ACTIVITY 2   DIFFUSION AND OSMOSIS

1. Work in groups. Get the following supplies: **one plastic jar, one piece of dialysis tubing, and two long pieces of thread.**

    Dialysis tubing is an example of a selectively permeable membrane. Cell membranes are also selectively permeable.

2. **Wet the piece of dialysis tubing** for a minute or two and open it as demonstrated by your instructor.

    Construct a **dialysis tubing bag. Fold about one inch** of the tubing at one end, as shown in **Figure 4-2. Tie it tightly** with the thread. The thread should be wrapped several times around the **doubled** end of the tubing. Cut off any excess string.

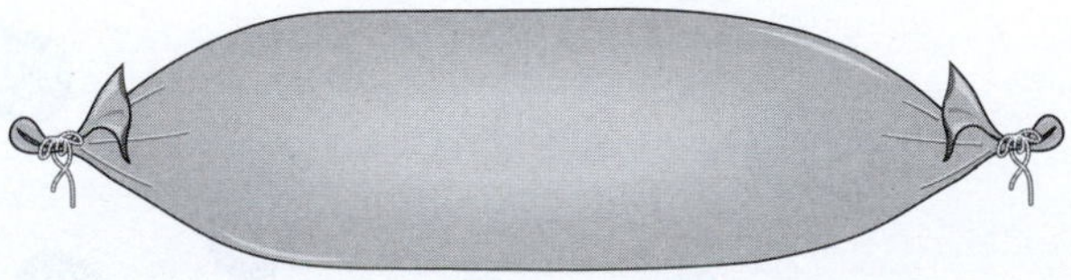

FIGURE 4-2.   Dialysis Tubing Bag

3. Get a bottle of **stock solution.** The solution consists of **water** with a mixture of dissolved **salt** (NaCl), **sugar** (glucose), and **starch.**

Hint:

Starch tends to settle out on the bottom of the bottle. Shake the stock solution to completely mix it before measuring.

---

4. Using a graduated cylinder, measure **20 milliliters (ml)** of stock solution and pour it into your dialysis tubing bag.

   Repeat the procedures outlined in step #2 above to seal the open end of the dialysis tubing bag. Cut off any excess string.

5. **Weigh your bag** according to the following directions, using the balance in your classroom:

   - Make sure the **balance is set to zero.** (If you have any problems using the scale, consult your laboratory instructor.)

   - Weigh the bag to **the nearest tenth of a gram** (for example, **19.4 g**)

   - Record the **initial weight** here: _____ g

6. Fill the plastic jar about three-quarters full with **distilled water.** Don't fill the jar with tap water. **Immerse** the dialysis tubing bag in the jar of distilled water.

7. **Write a hypothesis.** What do you think will happen to the **weight** of the bag during the 20 minutes it sits in the jar of distilled water?

   Hypothesis:

8. Set the jar aside for **at least 20 minutes.**

   While you're waiting, complete the Comprehension Check and then set up the controls for chemical testing in Activity 3.

   You'll return to your dialysis experiment when **at least 20 minutes** have elapsed.

FIGURE 4-3.   Dialysis Bag Surrounded by Distilled Water

✓ Comprehension Check

1.   When someone uses a 10% glucose solution, what is actually in this solution? Fill in the blanks with the correct percentages (this **must** add up to 100%).

   _____ % glucose _____ % water

2.   **Figure 4-3** represents a dialysis tubing bag containing dissolved salt, sugar, and starch, surrounded by distilled water.

   (Circle one answer.)

   The **highest** concentration of **salt** is **inside the bag / outside the bag.**

   The **highest** concentration of **sugar** is **inside the bag / outside the bag.**

   The **highest** concentration of **starch** is **inside the bag / outside the bag.**

   The **highest** concentration of **water** is **inside the bag / outside the bag.**

3.   Complete the following sentences.

   If my dialysis tubing bag **gains weight** during the experiment, I'll know that water _____ the bag.

   If my dialysis tubing bag **loses weight** during the experiment, I'll know that water _____ the bag.

Check your answers with your instructor before you continue.

ACTIVITY 3

CONTROLS FOR THE DIFFUSION AND OSMOSIS EXPERIMENT

The purpose of the dialysis tubing experiment is to determine which molecules are able to cross the selectively permeable membrane of the dialysis tubing bag. To determine whether starch, salt, and sugar are able to cross the membrane, you'll perform **three** different chemical tests. Each uses an **indicator chemical that changes color** in the presence of **one** of the following molecules: **starch, salt,** or **sugar.**

a.   **Iodine Test for the presence of starch.** When added to a solution, iodine will turn **black** if starch is present. If no starch is present, iodine will remain **reddish-brown** in color.

b.   **Silver Nitrate Test for the presence of chloride ions.** When added to a solution that contains **chloride ions,** silver nitrate will change color from **clear** to **cloudy white.** (Remember: salt is composed of sodium ions and chloride ions.)

c.   **Benedict's Test for simple sugars**. When added to a solution that contains simple sugars, such as glucose, Benedict's solution will change color from **turquoise blue** to one of the following colors: **green, yellow, orange**, or **red.** Green indicates the smallest amount of simple sugar and red indicates the highest.

Note:

It's important to remember that the starch, salt, and sugar are not changing in any way. It's the indicator chemicals that are changing color.

---

1.   Work in groups. Get a **large beaker.** Fill it **one-third full** with **tap water** and heat it to boiling.

   While you're waiting for the water to boil, set up the materials needed for chemical testing.

2.   To set up your **control** tubes, get the following supplies: **a test tube rack, a ruler, a marking pencil, a test tube holder, and three test tubes.**

   Label the test tubes salt, sugar, and starch.

3.   Using the ruler, place a **line** on each test tube **1 cm from the bottom** of the tube.

   Fill each of the three test tubes to the 1-cm line with stock solution.

4. To the test tube labeled **salt,** add **one dropper full** of **silver nitrate** solution.

Caution!

Avoid getting this solution on your skin.

---

5. Shake the tube gently to mix the contents. **Observe the color** of the solution. **Record your results** in the appropriate column of **Table 4-1.**

Note:

Save all three of your control test tubes (salt, starch, and sugar) for later use.

---

6. To the test tube labeled **starch,** add **one dropper full** of **iodine** solution. Shake the tube gently to mix the contents.

    Observe the color of the solution. **Record** your results in **Table 4-1.**

7. To the test tube labeled **sugar,** add **one dropper full** of **Benedict's** solution. Shake the tube gently to mix the contents.

8. **Carefully** lower the test tube into the boiling water bath and allow it to remain for about **two minutes.**

Caution!

Hot glass looks exactly like cold glass. Use a test tube holder to remove test tubes from the boiling water. Don't leave the test tube holder clamped on the test tube while in the boiling water. Hot metal looks like cold metal!

---

9. **Using a test tube holder,** remove the tube from the boiling water bath and place it in the test tube rack.

10. **Observe the color** of the solution. **Record** your results in **Table 4-1.**

11. **Circle one answer.**

If **starch** leaves the dialysis tubing bag, I expect that the **iodine** test results on the water **in the jar** will be **positive / negative.**

If **sugar doesn't** leave the dialysis tubing bag, I expect that the **Benedict's** test results on the water **in the jar** will be **positive / negative.**

If **salt** leaves the dialysis tubing bag, I expect that the **silver nitrate** test results on the water **in the jar** will be **positive / negative.**

Check your answers with your instructor before you continue.

TABLE 4-1 RESULTS OF CHEMICAL TESTS				
INDICATOR SOLUTION	SHOWS THE PRESENCE OF	INITIAL COLOR	FINAL COLOR	RESULTS (+ OR −)
Silver Nitrate Test Control tube				
Iodine Test Control tube				
Benedict's Test Control tube				
Silver Nitrate Test Experimental tube				
Iodine Test Experimental tube				
Benedict's Test Experimental tube				

ACTIVITY 4
COMPLETING THE DIFFUSION AND OSMOSIS EXPERIMENT

1. Remove your dialysis tubing bag from the jar of water to test for osmosis. Follow the weighing instructions in **Activity 2, step #5** to obtain the final weight of your bag.

2. Record the **final weight** here: _____ g

    Did the bag **gain** or **lose weight?** _____

    Was your hypothesis about the **weight change** correct? _____

3. Calculate the **percent change in weight** and record your answer.

    $$\text{percent change} = \frac{\text{final weight} - \text{initial weight}}{\text{initial weight}} \times 100$$

    percent weight change = _____ %

4. What do you think **moved through the membrane** to cause this result?

5. To test for **diffusion, get three more test tubes. These are the three experimental tubes. Label the test tubes salt, sugar, and starch.**

    Using the ruler, place a **line** on each test tube **1 cm from the bottom** of the tube.

    Fill each of the three test tubes to the **1-cm line** with **water from the plastic jar that held the dialysis tubing.**

    Test the water for salt, sugar, and starch using the same indicator chemicals you used when setting up the controls in **Activity 3.**

    Each test tube should contain only water from the jar the dialysis tube was soaking in and the indicator chemical.

    Don't add stock solution to these test tubes!

6. **Record** the results of all three tests in **Table 4-1.**

    According to your test results, **list the molecules** that were able to **diffuse** out of the dialysis bag.

7. If molecules are **expected** to move from **high concentration** to **low concentration,** why didn't **all** the molecules (salt, sugar, and starch) leave the dialysis tubing bag?

8. The ability to **separate molecules from each other** using a **selectively permeable** membrane is referred to as **dialysis.** Did the experiment you just finished illustrate this concept? **Explain** your answer.

Check your answers with your instructor before you continue.

ACTIVITY 5   OSMOSIS IN ELODEA CELLS

1. Work in groups. You'll be making **two different wet mounts** of elodea cells.

    One will be made with **culture water** from the elodea container. The other will be made by substituting **20% saline** for the water.

    Get the following supplies: **two elodea leaves, forceps, a dropper bottle of 20% saline solution, two slides, and two cover slips.**

2. Place a drop of 20% saline solution on the first slide and add an elodea leaf. Gently lower a cover slip over the specimen.

    With a marking pencil, write "S" on the right side of the slide to indicate saline solution.

3. Place a drop of culture water from the elodea container (not tap water) on the second slide and add an elodea leaf. Gently lower a cover slip over the specimen.

4. Place **two microscopes** side by side and **compare** the two slides. **Start with the 4× (scanning) lens.** Center and focus the specimen. Increase the magnification to **high** power.

5. Wait **five minutes,** and then check both slides again.

    Draw a large picture showing **one or two representative cells** from each slide.

ELODEA CELLS IN CULTURE WATER	ELODEA CELLS IN 20% SALINE SOLUTION

6. Add the following **labels** to your diagrams: **cell wall, cytoplasm,** and **chloroplasts.**

7. In **one** of your slides, you should be able to see the elodea **cell membrane.**

   On which slide is the cell membrane visible? _____

   Add a label for the cell membrane to your drawing.

8. Which slide represents the **control** in this experiment? **Explain** your answer.

9. What **process** caused the difference in appearance between the elodea cells in culture water and those exposed to the 20% saline solution?

Hint:

Don't forget that mature leaf cells contain a large central vacuole that functions in water storage.

Check your diagram labels and your answers with your instructor before you continue.

✓ Comprehension Check

1. Assume that the interior of an elodea cell is **98% water** molecules. The remaining **2%** is composed of various dissolved particles.

   The salt solution you applied to the slide is _____ % **water** molecules and _____ % **salt** particles.

2. Using the information in **question 1, add an arrow** to the **picture** you drew of the **elodea cells in saline solution,** showing the **direction of water movement.**

3. **(Circle one answer.)** The process that occurred to the elodea cells in salt solution was an example of **diffusion / osmosis.**

   Explain your answer.

4. A student performed the elodea cell experiment using **human blood cells.** When the student placed the blood cells in **salt water,** they shriveled up and died. When the student placed other blood cells in **distilled water,** they filled up, exploded, and died.

   What **cellular organelle present in leaf cells, but not in blood cells,** prevented the elodea cells from shriveling or exploding during your experiment?

Check your answers with your instructor before you continue.

SELF TEST

1.   In **Figure 4-4, draw arrows** showing the movement of the water molecules.

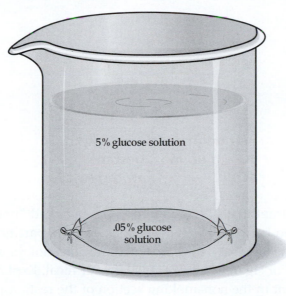

FIGURE 4-4.   Movement of Water Molecules

2.   A patient with partial kidney failure enters the hospital. The doctor tells the patient that the kidney is responsible for filtering and removing waste molecules from body fluids. The doctor recommends use of a **kidney dialysis** machine. Using your knowledge of **diffusion, osmosis,** and **dialysis,** what do you think the dialysis machine will do for this patient?

3.   You're admitted to the hospital with severe dehydration. A student nurse is directed to give you a transfusion of **blood plasma** to replace your lost body fluids. The nurse gives you a transfusion of **distilled water** by mistake. **Explain** what will happen to your **red blood cells.** In your explanation, use the following words: **cell membrane, osmosis, high concentration,** and **low concentration.**

4. Will a sugar cube dissolve faster in iced tea or hot tea? _____

    Explain your answer. In your explanation, use the following words: **diffusion, high concentration, low concentration, molecular movement,** and **collisions.**

5. You're in a restaurant that's divided into "smoking" and "nonsmoking" sections. There's a partition between the two sections, but the partition doesn't reach the ceiling. Despite the fact that you're seated in the nonsmoking section, you're bothered by the smell of cigarette smoke during your meal. Explain why the cigarette smoke was present in the nonsmoking section of the restaurant. In your explanation, use the following terms: **diffusion, molecules, higher concentration,** and **lower concentration.**

Investigating Cellular Respiration

Objectives

After completing this exercise, you should be able to:

- explain the process of aerobic cellular respiration in plants and animals
- summarize experimental results showing that microorganisms also perform aerobic cellular respiration
- compare the rate of aerobic cellular respiration and carbon dioxide production while at rest with that during exercise
- explain the similarities and differences between aerobic respiration and anaerobic respiration and give examples
- explain the similarities and differences between ethanol fermentation and lactic acid fermentation
- apply your knowledge of aerobic and anaerobic respiration to real-life situations

CONTENT FOCUS

Cellular respiration is the process by which living organisms convert the **chemical energy** in food (**organic molecules**) to a useful form for cellular functions. This conversion process is comparable to the production of electrical energy to run machines in your home. A television can't take energy directly from coal, gasoline, or even nuclear fuel. These fuels must be converted to electricity by your community power plant before they can be used by your home appliances.

Cells must transfer the energy stored in food molecules into **adenosine triphosphate (ATP),** a form of energy that can be used to run cell activities. If you read the ingredients on a candy wrapper or a box of corn flakes, however, you won't find ATP listed. The energy in the foods has to be **transferred** into ATP. This conversion process is called **aerobic cellular respiration.**

During **aerobic respiration,** living organisms extract energy from the chemical bonds in food molecules (such as carbohydrates, fats, or proteins) and convert that energy into ATP.

The process of cellular respiration can be summarized with this equation:

food molecules + oxygen → carbon dioxide + *ATP energy* + water + heat

Cellular respiration isn't the same as breathing. It occurs continuously, day and night, in all living cells. If it stops, cells will quickly die. Aerobic respiration is not a single chemical reaction, but it involves as many as 50 intermediate steps and the formation of a number of different compounds before ATP energy is finally produced. Specific enzymes are needed to control each step. For all living organisms, the chemical reactions of cell respiration are amazingly similar. Many pathways are virtually identical, even between such dissimilar organisms as bacteria and human beings.

The process of aerobic cellular respiration **uses oxygen** and **produces carbon dioxide as a waste product.** For this reason, cellular respiration can't take place without gas exchange (breathing). When you **inhale, oxygen is carried by the bloodstream** to all the cells of your body. As **blood circulates** through your tissues, it **picks up the carbon dioxide produced by cellular respiration** and transports it to the **lungs** for removal. Every time you **exhale, carbon dioxide** is released.

To sum up, aerobic respiration is a process that uses oxygen to burn the food (fuel) we consume and to produce useful ATP energy. The more energy your body needs, the more fuel and oxygen you must take in. It should now be clear why your breathing rate (and therefore your oxygen intake) increases when you exercise. During exercise, you also increase the amount of carbon dioxide you produce because you burn more food molecules to produce ATP.

ACTIVITY 1
DO MICROORGANISMS PERFORM CELLULAR RESPIRATION?

Yeasts are microscopic fungi that are commercially very important. They perform aerobic cellular respiration using pathways similar to those found in larger plants and animals.

Methylene blue dye can be used as an **indicator** for cellular respiration in yeast.

Aerobic respiration releases hydrogen ions and electrons that are picked up by the methylene blue dye, gradually turning the dye **colorless.** The mitochondria of yeast cells undergoing cellular respiration will appear as a clear area surrounded by a ring of light blue cytoplasm. The nucleus may be visible as a small, darkly stained spot.

If cellular respiration isn't taking place, the mitochondria will absorb the blue dye and will not turn colorless. The cells will appear to have a large, **darkly stained** central area surrounded by the ring of light blue cytoplasm.

1. Work in groups. Get the following supplies: **slides, cover slips, yeast suspension, a pipette, and a dropper bottle of methylene blue dye.**

2. Place a **drop of yeast suspension** on a clean microscope slide. Add **one small** drop of **methylene blue** dye and place a **cover slip** over the mixture. Observe the yeast cells with the **high-power objective.**

3. Is cellular respiration occurring in any of the yeast cells on your slide? _____

    What percentage of the yeast cells are **not** undergoing cellular respiration? _____
    (Make an estimate.)

    Why **isn't** cellular respiration taking place in all the cells?

4. **Draw one yeast cell** that is **undergoing cellular respiration** and **one that isn't.** Make the drawings **large and clear.** Label the **cytoplasm, nucleus (if visible),** and **mitochondria** of your yeast cells.

YEAST CELL CELL RESPIRATION OCCURRING	YEAST CELL CELL RESPIRATION ABSENT

Check your drawings with your instructor before you continue.

ACTIVITY 2
COMPARISON OF CLASSROOM AIR WITH EXHALED AIR

This experiment uses **limewater** (a very concentrated **calcium carbonate solution**) as an **indicator** solution. Limewater turns cloudy when CO_2 is added to it.

1. Work in groups. Get the following supplies: **two small beakers, one drinking straw, a pipette, and a container of limewater.**

2. **Fill** each beaker **about half full with limewater.** Label the beakers 1 and 2.

> ### Note:
>
> **Read through the instructions COMPLETELY before you continue.**

3. Do you think bubbling **room air into Beaker 1** will cause the limewater to turn cloudy? **Enter your hypothesis in Table 5-1.**

4. Do you think bubbling **your exhaled air into Beaker 2** will cause the limewater to turn cloudy? **Enter your hypothesis in Table 5-1.**

TABLE 5-1
EFFECT OF BUBBLED AIR ON LIMEWATER

	HYPOTHESIS (YES OR NO)	RESULTS
Will room air cause the limewater in Beaker 1 to turn cloudy?		
Will exhaled air cause the limewater in Beaker 2 to turn cloudy?		

5. To bubble **room air, repetitively squeeze and release** air from the **pipette** into the liquid in **Beaker 1.** Continue this process for **one minute.**

   Observe the beaker during the bubbling process and **record** your results in **Table 5-1.**

> ### Caution!
> **Be careful not to suck the solution into your mouth!**

6. To bubble test your **exhaled air,** blow **very gently** through the drinking straw into the liquid in **Beaker 2.** Continue this process for **one minute.**

    Observe the beaker during the bubbling process and record your results in **Table 5-1.**

7. When you've completed your experiment, empty your beakers into the **waste container.**

✔ Comprehension Check

1. In one or two sentences, **summarize the results** of this experiment.

2. On the basis of your results, **what gas was bubbled** into Beaker 2? _____

3. In reference to **Question 2,** what **process** produced this gas? _____

4. If people in the classroom are exhaling, why do you think your Beaker 1 results were negative?

Check your answers with your instructor before you continue.

ACTIVITY 3   EFFECT OF EXERCISE ON CARBON DIOXIDE PRODUCTION

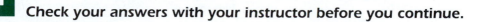

Breathing is controlled by a "breathing center" in the medulla of the brain. This center is activated by a **buildup of carbon dioxide in the blood or a low oxygen concentration,** such as occurs at high elevations. This experiment will compare the **amount** of CO_2 produced during exercise with that produced at rest.

We'll use **bromothymol blue** as an **indicator** that will tell us **how much** CO_2 is produced. Recall that bromothymol blue turns **green or yellow** when CO_2 is **added** and returns to **blue** when CO_2 is **removed.**

1. Work in groups. Get the following supplies: **three 400-ml beakers, bromothymol blue solution, ammonia solution, a plastic bag, a coupler tube, a mouthpiece, a rubber band, and an air stone assembly.**

2. Fill each beaker **with 200 ml of bromothymol blue indicator solution. Label** the beakers **1, 2, and 3.**

3. **Attach the coupler tube** to the plastic bag and hold the tube in place with the rubber bands (as shown in **Figure 5-1**).

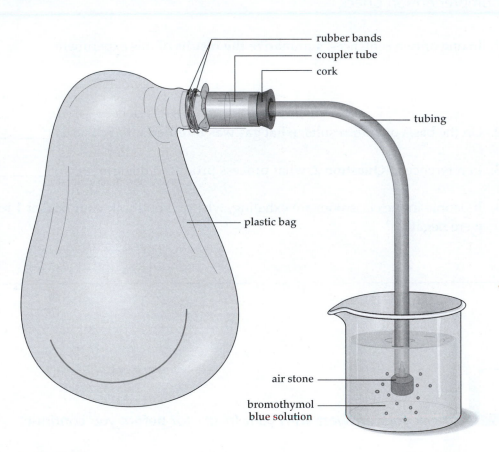

FIGURE 5-1.   Apparatus to Bubble Captured Air into Bromothymol Blue Solution

Testing Room Air

1. Fill the plastic bag with **room air** by pulling it through the air ("rounding out" the bag).

    Quickly plug the bag with the **air stone assembly. Be careful not to let the air escape.**

2. Place the **air stone** into the indicator solution in **Beaker 1** (see **Figure 5-1**).

    Squeeze the air out of the plastic bag so that it bubbles into the beaker.

Ammonia Treatment

1. Using the dropper bottle, add **ammonia** to **the beaker one drop at a time** and **gently** stir after adding each drop.

   Be sure to **keep an accurate count** of the number of drops added.

2. Continue adding ammonia until the indicator solution changes to a **light green color and remains green for 30 seconds.**

3. **Record** the number of drops you added in **Table 5-2.**

   The amount of ammonia needed to change the indicator solution from **blue to green** will be used as an estimate of **the amount of CO_2 present** in the room air.

T A B L E 5 - 2		
EFFECT OF EXERCISE ON CO_2 PRODUCTION		
SAMPLE	COLOR AFTER BUBBLING	DROPS OF AMMONIA USED
Room air (Beaker 1)		
Exhaled air at rest (Beaker 2)		
Exhaled air after exercise (Beaker 3)		

Testing Exhaled Air Before Exercise (at Rest)

1. Secure the mouthpiece onto the plastic bag and fill the bag with exhaled air, crimping the end of the bag with your hand to prevent the air from escaping. **Quickly** plug the bag with the **air stone assembly,** being careful not to lose the air.

2. Place the air stone into the indicator solution in **Beaker 2.** Squeeze the air out of the plastic bag so that it bubbles into the beaker. Keep squeezing until the plastic bag is empty.

   Add ammonia one drop at a time until the indicator solution changes to a **light green color and remains green for 30 seconds. Record** the number of drops you added in **Table 5-2.**

Testing Exhaled Air After Exercise

1. Secure the mouthpiece on the plastic bag. Set the bag close at hand and **exercise vigorously for two minutes.** You may jog around the room, do jumping jacks, or step rapidly up and down on an aerobic step.

   Fill the plastic bag with exhaled air **immediately** and plug the bag with the **air stone assembly.**

2. Bubble the exhaled air into **Beaker 3** as you did when testing the exhaled air before exercise. **Add ammonia** until the indicator solution changes to a **light green color and remains green for 30 seconds. Record** the number of drops you added in Table 5-2.

Comprehension Check

1. What happened to the amount of CO_2 in your exhaled air after exercise? **Explain** your answer.

2. How do you think it would affect the amount of CO_2 produced if you exercised for ten minutes instead of only two minutes? **Explain** your answer.

3. When you exercise, your heart beats faster and the blood carries more oxygen to your muscles. What do muscle cells do with this oxygen?

Check your answers with your instructor before you continue.

ACTIVITY 4   ALTERNATIVES TO AEROBIC RESPIRATION

You've demonstrated that cells can carry out aerobic respiration when oxygen is available, but they can also produce ATP energy by an anaerobic process called **fermentation** when oxygen is in short supply. The fermentation pathway can produce a variety of end products (refer to **Figure 5-2**).

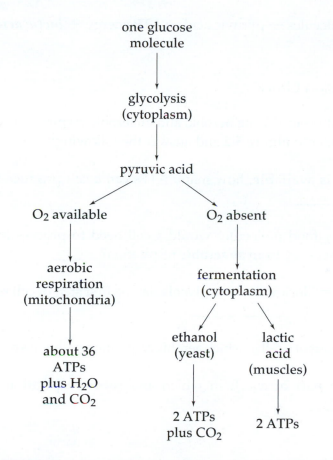

FIGURE 5-2.   Steps of Aerobic and Anaerobic Cellular Respiration

In most plant cells and in yeasts, pyruvic acid is converted to ethanol (drinking alcohol) and carbon dioxide. Alcohol fermentation by yeast cells is used commercially to manufacture baked goods and alcoholic beverages.

Alcohol fermentation is summarized with the following equation:

food molecules → pyruvic acid → carbon dioxide + ATP energy + *ethanol* + heat

Another method of anaerobic respiration is called **lactic acid fermentation.** During lactic acid fermentation, the pyruvic acid molecules from glycolysis are converted into **lactic acid.**

Some bacteria specialize in lactic acid fermentation. The lactic acid produced by bacteria causes milk to sour. A similar process occurs when cabbage ferments to form sauerkraut. The dairy industry uses lactic acid fermentation by bacteria and fungi to produce sour cream, cheese, and yogurt. Your muscle cells may use lactic acid fermentation when oxygen demands can't keep up with supply (such as during vigorous exercise).

The equation for lactic acid fermentation is

food molecules → pyruvic acid → ATP energy + *lactic acid* + heat

✔ Comprehension Check

The steps that occur during aerobic and anaerobic respiration are summarized in **Figure 5-2**. Refer to **Figure 5-2** and answer the following:

1. If **oxygen is available,** how many ATPs can a cell produce from **one** glucose molecule? _____

2. How many food molecules would a cell need to process to obtain the same amount of energy from **anaerobic** respiration? _____

3. Name the cell location where **glycolysis** (the first step of cell respiration) occurs.
   _____

4. Name the part of a cell in which **aerobic** respiration pathways occur. _____

5. Name the part of a cell in which **anaerobic** respiration pathways occur.
   _____

Check your answers with your instructor before you continue.

ACTIVITY 5   ANAEROBIC RESPIRATION

Note:
Read through the instructions COMPLETELY before you begin.

1. Work in groups. Get the following supplies: **a pair of safety glasses, a tablespoon, two 400-ml beakers, a container of limewater, and a jar of sucrose.**

    In your classroom, you'll find two **flasks** on a **warming plate.** The flasks contain **water** and **yeast cells.**

    Each flask is closed with a stopper holding two pieces of tubing (see **Figure 5-3**). The longer piece of tubing is connected to an **air stone.**

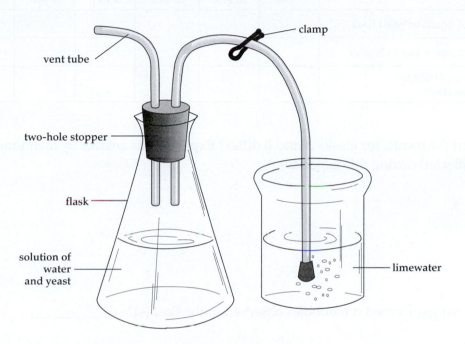

FIGURE 5-3. Setup for a Fermentation Experiment

2. Fill both beakers **half full** with **limewater.** Place the beakers on the table next to the warming plate. Insert the air stone from **Flask A into the first beaker** and from **Flask B into the second beaker.**

Caution!

Eye protection is required for the following procedures!

---

3. **Wearing eye protection,** gently remove the stopper from **Flask B and add one tablespoon of sucrose.**

    Replace the stopper. **Record the time** when you added the sucrose. _____

4. **Immediately** after adding the sucrose, remove the clamp from the air stone tube and **switch it to the vent tube. Switch the clamps on Flask A and Flask B.**

5. Check on your experiment every **five minutes** for the next **15 minutes. Record** the results for **each five-minute period** in **Table 5-3.**

T A B L E 5 - 3 **OBSERVATIONS OF YEAST FLASKS**						
OBSERVATIONS	EXPERIMENTAL FLASK A (WATER AND YEAST)			EXPERIMENTAL FLASK B (WATER, YEAST, AND SUCROSE)		
	5 MIN.	10 MIN.	15 MIN.	5 MIN.	10 MIN.	15 MIN.
Color of liquid in yeast flask						
Color of limewater in beaker						
Presence of bubbles in yeast flask						

6. Did the results for Flasks A and B differ? **Explain** your answer by mentioning facts collected during this experiment.

7. What gas formed the bubbles observed in the flask(s)? _____

    Explain your answer.

Caution!

Eye protection is required for the following procedures!

8. **Wearing eye protection,** carefully remove the stopper from **Flask A** and sniff the released air. **Repeat** the process with **Flask B. Describe** the results.

✓ Comprehension Check

1. A yeast cell undergoing fermentation produced **six ATP molecules.** How many glucose molecules were used?_____ **Explain** your answer.

2. If the same cell was undergoing aerobic cellular respiration instead of fermentation, but using the same number of glucose molecules, how many ATP molecules would be produced?_____ **Explain** your answer.

Check your answers with your instructor before you continue.

Note:
Lactic acid fermentation can satisfy only the body's short-term energy needs. If the oxygen shortage continues too long, lactic acid accumulates in your muscle tissue, lowering the pH. Eventually, this acidic environment inhibits the activity of enzymes that are needed for muscle function. Muscle fibers lose their ability to contract, a process called muscle fatigue.

   The accumulated lactic acid can damage muscle proteins, producing the soreness and pain often associated with lengthy, strenuous exercise.

   When oxygen is available again, the lactic acid is converted back to pyruvic acid and aerobic respiration can continue as before.

SELF TEST

1. You and several other students observe that dead animals along the roadside often increase in size several days after they have been killed. In the spirit of scientific investigation, you collect the gas from inside one of these dead animals and bubble it through limewater. The limewater turns cloudy.

    a. What does the cloudy result indicate?

    b. If the animal is dead, how are these results possible? **Explain your answer in detail.**

2. Which of the following produces the most energy in the form of ATP?

    a. aerobic respiration
    b. anaerobic respiration
    c. alcohol fermentation
    d. all of the above produce the same amount of ATP energy, but through different chemical reactions
    e. none of the above produces ATP

3. **Bromothymol blue** is an indicator solution that **turns green or yellow** when CO_2 is added and **turns back to blue** when CO_2 is removed. Carbon dioxide is bubbled into a solution of bromothymol blue until the solution turns yellow. A sprig of elodea (an aquatic plant) is placed in the yellow solution. After a few hours in the sunlight, the yellow solution turns blue again. **Suggest an explanation** for the color changes observed in the bromothymol blue indicator.

For Questions 4 through 7, use the following three answers. An answer may be used once, more than once, or not at all. For each question, explain your answer.

  a. process occurs only under aerobic conditions
  b. process occurs only under anaerobic conditions
  c. process occurs under both aerobic and anaerobic conditions

4. _____ Manufacturing beer or wine

   Explanation:

5. _____ ATP production

   Explanation:

6. _____ Accumulation of lactic acid in a runner's muscles causes her to drop out of the race.

   Explanation:

7. _____ Production of 36 ATPs, water, and carbon dioxide from one glucose molecule

   Explanation:

8. If an elephant's cells were to lose their ability to perform aerobic cellular respiration, which of the following would most likely happen?

   a. the cells would switch over to anaerobic respiration, and the elephant would be fine

   b. the cells would switch over to lactic acid fermentation, and the elephant would survive but would have achy muscles

   c. the cells would switch over to alcohol fermentation, and the elephant would survive but would be in a drunken stupor

   d. the cells would die because the elephant's energy needs can't be met using anaerobic metabolic pathways

   e. the cells could live on stored energy for an extended period, and cellular respiration wouldn't be necessary for several years

   Explain your answer.

Photosynthesis

Objectives

After completing this exercise, you should be able to:

- list each of the four components required for photosynthesis to occur and the two end products that are produced
- use a pH indicator to demonstrate the uptake of carbon dioxide and release of oxygen during photosynthesis
- discuss the connection between starch production and the process of photosynthesis
- examine the physical features of leaves that function in gas exchange during photosynthesis

CONTENT FOCUS

Most of the energy that fuels the living systems on earth comes from the sun. The sun's energy is captured by the process of photosynthesis. Photosynthesis occurs in both plants and microorganisms, although the process may be slightly different from the equation described here.

The process of photosynthesis can be summarized with the following equation:

$$CO_2 + H_2O \xrightarrow[\text{chlorophyll}]{\text{sun's energy}} C_6H_{12}O_6 + O_2$$

carbon dioxide   water   glucose   oxygen

Photosynthesis makes use of light energy to manufacture food molecules from carbon dioxide and water. Photosynthesis is the most important chemical process on earth. Photosynthetic organisms have the green pigment **chlorophyll,** which captures the light energy used in photosynthesis.

During photosynthesis, carbon dioxide is removed from the atmosphere and used to form **glucose.**

Four "ingredients" are needed for photosynthesis to take place:

- **Sunlight**—provides the energy used to form glucose. **Glucose** (a sugar) is the end product of photosynthesis. The process of photosynthesis **converts solar energy into food energy** (the light energy from the sun is converted into chemical energy and stored in the chemical bonds of glucose molecules).
- **Chlorophyll**—producers trap light energy with the pigment chlorophyll. (Note: ONLY organisms that have chlorophyll can perform photosynthesis. Some organisms have additional photosynthetic pigments, but chlorophyll is always present.)
- **Carbon dioxide**—provides the carbon and oxygen needed to build glucose molecules.
- **Water**—provides the hydrogen and oxygen needed to build glucose molecules.

So, to summarize the process, plants and photosynthetic algae remove carbon, hydrogen, and oxygen atoms from the air or water and use these ingredients to construct glucose molecules. The energy in sugar and other food molecules originates as solar energy and is converted to food energy through photosynthesis.

Oxygen is a waste product of most photosynthesis. Photosynthesis provides the oxygen in our atmosphere (the oxygen you breathe every day). The atmosphere of the early planet earth, before there were photosynthetic organisms, did not contain any free oxygen. Our current atmosphere is about 21% oxygen, which has accumulated from millions of years of photosynthesis.

ACTIVITY 1   ABSORPTION OF CARBON DIOXIDE DURING PHOTOSYNTHESIS

This experiment will be conducted with elodea, an aquatic plant (photosynthesis occurs in the water as well as on land). The process of photosynthesis removes carbon dioxide from the water as photosynthesis continues, reducing the carbon dioxide levels in the immediate environment.

When CO_2 is combined with water, it forms carbonic acid, resulting in an acidic solution. As a hypothesis, therefore, it would be logical to predict that as a plant takes up CO_2 to use for photosynthesis, the pH of the surrounding water would gradually become more basic.

To test this hypothesis, you'll use a pH indicator called phenol red. Phenol red turns yellow in an acidic solution and red in a basic solution.

1. Work in groups. From the supply area, get **two large test tubes, a test tube rack, two test tube stoppers, one healthy sprig of elodea, a drinking straw, a 1,000-ml beaker, a 400-ml beaker, a goose-neck lamp, a glass petri dish, a scalpel (or razor blade), and a dropper bottle of phenol red indicator solution.**

2. Pour **100 ml of tap water** into the **400-ml beaker;** add **15 drops** of **phenol red indicator.** Place the drinking straw in the beaker.

Blow through the drinking straw until the phenol red indicator changes color to **yellow (or orange-yellow).**

3. Set both test tubes into the rack and pour 50 ml of the acidified phenol red solution into each test tube. Place the stopper into one of the test tubes.

4. Fill the petri dish with tap water. Place the stem end of the elodea sprig into the water in the petri dish.

    Using a sharp scalpel or razor blade, quickly make a diagonal cut across the base of the stem (while holding the stem under the surface of the water).

    Don't let the cut end of the elodea dry out! Place the plant into the test tube **immediately** after you make the cut across the stem—**stem side down.** Insert the stopper. **The elodea sprig needs to be completely covered with water.**

5. Fill the 1000-ml beaker about 3/4 full of tap water. Set up the lamp and turn on the light. The light should be directed toward the rack with the two test tubes.

    Place the beaker of water between the lamp and the test tube rack. The water will form a heat barrier between the lamp and the elodea.

    The placement for all components of this experiment is shown in **Figure 6-1.**

FIGURE 6-1.   Setup for CO_2 Absorption Experiment

6. Check the color of the solutions in the two tubes every 15 minutes for one hour and observe the surface of the leaves for the presence of bubbles.

Record your results in **Table 6-1.**

Note:

To check the color of the solution, remove a tube from the rack and hold it vertically against a white background. Then place the tube back in the rack in front of the light source.

TABLE 6-1

RESULTS OF CARBON DIOXIDE ABSORPTION EXPERIMENT

ELAPSED TIME	OBSERVATIONS (COLOR OF SOLUTION AND PRESENCE OF BUBBLES)
15 min	
30 min	
45 min	
60 min	

✔ Comprehension Check

1. Which test tube was the control? _____

2. Why was a control tube needed in this experiment? Explain your answer.

3. How was the process of photosynthesis related to the production of bubbles and what gas was in the bubbles?

4. Is the color change observed in the test tubes related to the pH of the solution in each tube? Explain your answer.

5. What gas was removed from the experimental tube that caused the solution to change color? _____

    If the gas was removed from the solution, where did it go?

6. Instead of the experiment you just completed, consider the following experimental setup: two tubes, both with elodea, both placed in acidified phenol red solution. One tube was kept in the dark for 24 hours and the other was exposed to sunlight for the same period. What effect would this have on the color of the phenol red in both tubes? Explain your answer.

7. **Challenge Question!** If you took a sample of water from a lake affected by acid rain and repeated the elodea experiment using that lake water, would the difference in the water affect the length of time it would take to see a color change in the phenol red indicator solution? Explain your answer.

Check your answers with your instructor before you continue.

ACTIVITY 2   TESTING LEAVES FOR STARCH

What does a plant use for energy at night? During the winter? Just as you consume more calories than you use immediately and store the remainder as fat, plants store excess energy from photosynthesis for later use by **converting glucose into starch.**

The presence of starch in leaves, therefore, provides indirect evidence that photosynthesis has occurred in those leaves.

1. Work in groups. Get the following supplies: **one or two solid green leaves, one or two variegated leaves, a dropper bottle of iodine, two petri dishes, and some paper towels.**

    On your laboratory table, you'll also find **an electric hot plate, two 400-ml beakers, a pair of long forceps, two pairs of small forceps, safety glasses, and a large beaker.**

    > ### Note:
    > **More than one group will be using a hot plate setup at the same time, so coordinate your activities accordingly.**

2. Pour **300 ml of tap water** into the large beaker and set the hot plate on **high.** Heat the water to **boiling.**

    Boiling the leaves **softens the supporting walls of the leaf cells** so that chemical testing of the cell contents will be possible.

3. Fill **one of the 400-ml beakers** with **250 ml of tap water** and set it aside for use in **step #11** of the procedure.

4. In the appropriate boxes in **Table 6-2, draw two outlines of the solid green leaf and two outlines of the variegated leaf.** For each leaf, draw **one outline** in the "**before**" box and another outline in the "**after**" box.

    In the **before** box that has the outline of the variegated leaf, shade in **only** those areas that are colored **green.**

5. Using the long forceps, insert your leaves into the boiling water. Add the leaves from **all other groups** that are sharing the hot plate setup. **Boil the leaves for five minutes.**

6. While the leaves are boiling, go to the supply area and pour **125 ml of ethyl alcohol** into the **second 400-ml beaker.**

TABLE 6-2	
LEAVES BEFORE AND AFTER IODINE TEST	
OUTLINE OF SOLID LEAF BEFORE IODINE TEST	PRESENCE OF STARCH IN THE SOLID LEAF AS SHOWN BY THE IODINE TEST

OUTLINE AND COLOR PATTERN OF VARIEGATED LEAF BEFORE IODINE TEST	PRESENCE OF STARCH IN THE VARIEGATED LEAF AS SHOWN BY THE IODINE TEST

7. **Using the long forceps, and being careful not to burn yourself,** remove the leaves from the boiling water. Place **all the boiled leaves** into the beaker of alcohol.

Caution!
Eye protection is required for the following procedures!

8. **Wearing eye protection, carefully** place the beaker of alcohol into the larger beaker of boiling water (see **Figure 6-2**).

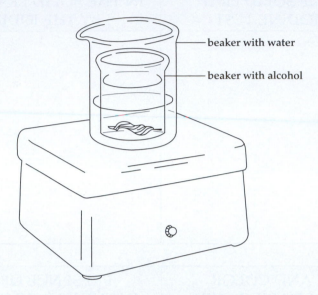

       beaker with water

       beaker with alcohol

FIGURE 6-2.   Setup for the Starch Experiment

9. Boil the leaves in the alcohol for about **five minutes,** or until the green color disappears and the leaves appear white. **Turn the hot plate off.**

10. Place a **small pool of iodine into each of the two petri dishes. Each group** that contributed leaves to the hot plate setup will need two **petri dishes with iodine.**

11. **Wearing eye protection, use the long forceps to carefully** remove the leaves, one at a time, from the alcohol.

    Dip one leaf at a time into the **beaker of tap water prepared in step #3** to remove excess alcohol, and gently place each leaf into the iodine in one of the petri dishes. If the leaves are wrinkled, use two pairs of small forceps to gently straighten them back to their original outlines.

12. **Soak the leaves** with iodine (see **Figure 6-3**). Wait for **two minutes.** Place the petri dishes on a **piece of white paper** and observe the leaves.

    Return to the leaf outlines you placed in the "after" boxes in **step #4** and **color in the starch pattern** that you observed.

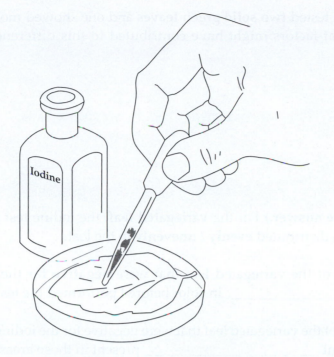

FIGURE 6-3. Starch Testing

✔ Comprehension Check

1. A positive iodine test shows that _____ is present in a leaf.

2. A positive iodine test indicates that the process of _____ has taken place in that leaf.

3. Rewrite the following false statement to make it true. It isn't necessary to keep the words in the same order.

   The starch in a leaf changes color when mixed with iodine.

4. **List the four ingredients** that are **necessary for photosynthesis** to take place.

   a.   c.

   b.   d.

5. **(Circle one answer.)** For the **solid green leaf,** my iodine test results were **positive / negative.**

   Approximately what percentage of your leaf contained starch? _____

6. If you had tested **two solid green leaves** and one showed more starch than the other, what factors might have contributed to this difference? **Explain** your answer.

7. **(Circle one answer.)** For the **variegated leaf,** the iodine test results show that starch was distributed **evenly / unevenly** in the leaf.

    The areas of the variegated leaf that were **negative** for the iodine test were _____ in color before I performed the test.

    The areas of the variegated leaf that were negative for the iodine test did not have the pigment _____ present in those areas.

8. Were all areas of the variegated leaf able to absorb equal amounts of sunlight? **Explain** your answer.

9. Suppose you placed a plant that contained starch in the dark and left it there for a week. At the end of the week, what results would you expect an iodine test to show? **Explain** your answer.

Check your answers with your instructor before you continue.

ACTIVITY 3   CELLS OF THE LEAF EPIDERMIS

Leaves are the primary locations of photosynthesis in most plants. The surface of the leaf is covered with small openings (pores) called **stomata** that allow gas exchange. Stomata can open and close.

Carbon dioxide, which is needed for photosynthesis to occur, enters through the stomata. Oxygen, a waste product of photosynthesis, escapes. Water is also lost through open stomata. The process is called **transpiration.**

Because most of the water lost by a plant escapes through the stomata, a plant closes its leaf stomata to prevent dehydration. When the stomata are closed, however, gas exchange can't occur.

In this activity, you'll observe the cells that form the stomata and other epidermal cells.

1. Work in groups. Obtain **a leaf, a compound microscope, forceps, a scalpel, a dropper bottle of distilled water, slides, and cover slips.**

2. Directions to make a wet mount of the leaf:

    ▪ Place a drop of distilled water onto a clean slide.

    ▪ **Gently** bend the leaf **toward the underside** until the **top cracks** but doesn't **completely break apart.** When the top cracks, you should see a **thin, transparent film** on the underside of the leaf. This is the **lower epidermis of the leaf. Remove it** and place it into the drop of **distilled water** on your slide.

    ▪ Place a cover slip on the slide.

3. Observe the epidermis on low power, and then change to a higher magnification. Note that several types of cells are present. Using **Figure 6-4** as a guide, locate the **epidermal cells** and the **guard cells** that **form the stomata.**

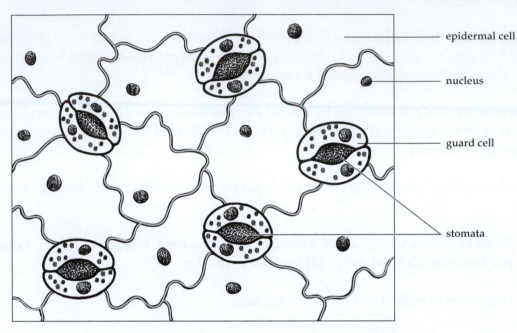

FIGURE 6-4.   Lower Epidermis and Stomata

4. As you look through the microscope at the leaf epidermis, which cells contain chloroplasts?

   How did you determine which cells contained chloroplasts?

5. In most leaves, there are many **more stomata on the lower epidermis than on the upper epidermis.** What is a possible benefit of this arrangement?

SELF TEST

1.   List three examples of organisms that perform photosynthesis that were **not** mentioned in this laboratory exercise:

2.   Give an example of a part of a plant in which **starch is stored,** but where photosynthesis **doesn't** occur: _____

3.   Based on your examination of stomata in the leaf epidermis, would you expect the leaves of a desert plant to have the same number of stomata as the leaves of plants found in Hawaii? Explain your answer.

4.   Considering your experiment with the phenol red indicator solution, what could you change in the experiment to increase the number of oxygen bubbles released from the elodea leaves? Explain your logic.

5.   Again, considering your experiment with the phenol red indicator solution, would you expect the rate of photosynthesis to remain constant in the experimental tube over a 24-hour period (with light provided for the entire 24 hours)? Explain your reasoning.

6.   At the following times of day, would you expect most of the stomata in a plant in your front yard to be opened or closed? Explain each answer.

     a. on a moist, cool morning

     b. on a hot, dry afternoon

7.   **Challenge Question!** A student did an experiment to test the hypothesis that "plants can't grow in cold temperatures." The student used two groups of identical plants, with 10 plants in each group. One group was placed in a **refrigerator** and the other on the **windowsill** in the classroom. Both groups were planted in the same type of soil and received exactly the same amount of water and fertilizer.

     The student checked on the plants twice a week for four weeks. At the end of the four weeks, the refrigerated plants were all dead, but the plants on the windowsill were fine. From the results of this experiment, the student concluded that the cold killed the plants.

     Do you agree with this conclusion? Explain your answer and support your explanation with facts about the experimental design.

EXERCISE 7

Organic Molecules and Nutrition

Objectives

After completing this exercise, you should be able to:

- test an unknown sample for the presence of organic molecules such as sugars, starch, protein, and lipids using indicator chemicals
- relate the nutrient content of a food to its original function in plants and animals
- discuss the benefits and drawbacks of using carbohydrates versus lipids as energy storage molecules for embryos
- use the Federal Dietary Guidelines to analyze the nutritional value of your meals

CONTENT FOCUS

All living organisms are composed of various types of **organic molecules** such as carbohydrates, lipids, proteins, and nucleic acids that make up their body tissues. Each of these four types of organic molecules (carbon-based compounds) has a different function in the body. Just as different parts of your body have different structures to support their functions, the same is true of other animals and plants. The **types** of **organic chemicals present** in different body tissues are related to the **function** of those tissues.

In this exercise, you'll be analyzing various food samples to determine which organic compounds are present. Carbohydrates, lipids, proteins, and nucleic acids **can't be seen directly. Indicator chemicals** can show you if these organic compounds are present in a food sample by **changing color.** Each type of indicator chemical is specific for one type of organic compound.

Analyzing foods for organic compounds using indicator chemicals can present some problems. Often, the **food itself has colors** that may interfere with the test results. It's easier to determine if an appropriate color change has occurred if you have a **standard for comparison.** Therefore, we must first establish a set of **positive and negative standards (controls)** for comparison in later experiments.

ACTIVITY 1   POSITIVE AND NEGATIVE INDICATOR TESTS

1. Work in groups. Get the following supplies: **a metric ruler, a marking pencil, six test tubes, a test tube rack, a test tube holder, and a large beaker.**

2. Using the ruler, place a **line** on each test tube **1 cm from the bottom** of the tube. **Number each test tube 1 through 6.** Place the marked test tubes in the test tube rack.

Benedict's Test for Simple Sugars

1. The chemical indicator in Benedict's solution reacts with **monosaccharides and some disaccharides;** however, for a reaction to occur **you have to heat the Benedict's solution.** Place the large beaker, half filled with tap water, on a hot plate. Set the temperature control to high until the water boils.

2. Fill **test tube 1** up to the line with **glucose** solution. (**Note:** Glucose is a monosaccharide, so it can be used as a **positive control** for Benedict's test.)

    Fill **test tube 2** up to the line with **distilled water.**

3. Fill the dropper with Benedict's solution. What color is Benedict's solution?
    _____

4. To **test tubes 1 and 2,** add **one dropper full of Benedict's solution.**

    Shake the test tubes to mix the solution.

5. **Using a test tube holder, carefully** place the tubes into the boiling water for two to five minutes.

Caution!

Hot glass looks exactly like cold glass! Use the test tube holder to remove test tubes from the boiling water!

Don't leave the test tube holder clamped on the test tube while in the boiling water. Hot metal looks exactly like cold metal!

6. **Using the test tube holder,** remove the tubes and **observe** the color in each test tube.

    If you see a color change ranging from **green through yellow, orange, or red,** this is a positive test for simple sugars.

    If the color in the tube **hasn't changed,** this is a **negative** test for simple sugars.

> ### Note:
> **Save tubes 1 and 2 for use in later experiments.**

✔ Comprehension Check

1. Which tube was positive for sugar? _____ Which was negative _____?

2. Based on the **results of Benedict's test, glucose** could be correctly identified as belonging to which of the following groups?

    a. lipids   d. simple sugars
    b. starches   e. choices b and d are both correct
    c. proteins

Iodine Test for Starch

1. Fill test tube 3 up to the line with starch solution. Fill test tube 4 up to the line with distilled water.

2. Fill the **dropper with iodine.** What color is iodine? _____

3. To **both tubes,** add one **dropper full of iodine. Shake** the test tubes to mix the solution.

4. Observe the color in each test tube. If the iodine changes color to black, that's a positive result for starch.

5. Which tube was positive for starch? _____ Which was negative _____?

6. Place **one small drop of iodine** in the box.

    On the basis of the results of the iodine test, what can you **conclude about the paper** used for this laboratory manual?

> ### Note:
> Save tubes 3 and 4 for use in later experiments.

Biuret Test for Protein

1. Fill **test tube 5** up to the line with **albumin (egg white)** solution. (**Note:** Egg white is mostly protein, so it can be used as a **positive control** for the Biuret test.)

    Fill **test tube 6** up to the line with **distilled water.**

2. To **tubes 5 and 6,** add one dropper full of **Biuret solution. Shake** the tubes to mix the solution.

3. Observe the color in each test tube. If the Biuret solution changes color from blue to purple, that's a positive result for protein.

4. Which tube was positive for protein? _____ Which was negative _____?

> ### Note:
> Save tubes 5 and 6 for use in later experiments.

Sudan IV Test for Lipids

1. Get the following supplies: **one piece of filter paper, forceps, a petri dish, and dropper bottles of vegetable oil, distilled water, and Sudan IV stain.**

2. Place the filter paper on a **clean** piece of paper (**not** on the laboratory counter, which may be dirty).

    Using a **pencil** (**not** the wax pencil), draw **two** circles about the size of a dime, spaced **apart** on the filter paper. Label the first circle **"oil"** and the second circle **"water."**

3. Place **one drop of oil** into the circle labeled **"oil"** and **one drop of distilled water** into the circle labeled **"water."**

4. **Set the filter paper aside to dry.** If needed, use a hair dryer **on low power** to evaporate excess liquid.

5. When the filter paper is **completely dry,** place it in the **petri dish** and cover the paper with **Sudan IV solution.** Let the paper soak in the stain for **three minutes.**

6. While the paper is soaking, get a **glass bowl** from the **supply area** and **fill it** with **distilled water.**

7.   When three minutes are up, place the filter paper into the bowl of distilled water. **Rinse gently** for **one minute.**

     Remove the filter paper from the water with the forceps and **observe the color of the two circles.**

8.   **A dark pink spot** indicates a **positive** test for lipids.

     A **pale pink** color, **no different from the rest of the paper,** should be considered **negative.**

9.   **Record** the results of your test (positive or negative) below. Enter a " **+** " if the test was **positive** and a " **−** " if the test was **negative.**

     "oil" circle _____   "water" circle _____

Note:
Save your test paper for use in later experiments.

✔ Comprehension Check

Using the results from the tests you've just completed, fill in **Table 7-1.**

T A B L E   7 - 1
INDICATOR TESTS

Indicator Test	Tests for	Negative Result (Color)	Positive Result (Color)
			green through red
	starch		
Sudan IV test			
Biuret test			
		light blue	

Check your answers with your instructor before you continue.

ACTIVITY 2   TESTING FOOD SAMPLES

Have you ever wondered what's really in the foods you choose? Is a food high in fat or protein? Is it a good source of carbohydrates? The answers to these questions become clearer when food samples are analyzed to determine which organic compounds are present.

   In the next activity, you'll analyze several commonly eaten foods and determine their nutrient content. Solid food samples have been blended into liquid, for easier testing.

1. For each of the foods listed in **Table 7-2,** form a **hypothesis** about which organic compounds it will contain.

    Record your hypotheses in **Table 7-2.**

    For each hypothesis, enter a "**+**" if you think the test will be **positive** and a "**−**" if you think the test will be **negative.**

TABLE 7-2 **HYPOTHESES FOR FOOD EXPERIMENTS**				
FOOD	SIMPLE SUGARS	STARCH	PROTEIN	LIPIDS
Lettuce				
Hamburger				
Tuna				
Milk				
Refried beans				
Peanut butter				

2. Work in groups. Your instructor will assign several foods for your group to analyze.

    Your group will **share its data** with other groups that are analyzing different foods.

3. For each food sample to be analyzed, get **three clean test tubes.** You'll also need **one piece of filter paper.**

4. Following the **test procedures exactly as you did in Activity 1,** analyze each food sample for the presence of **simple sugars, starch, protein, and lipids.**

5. **Compare** your test results with the **controls** you **saved from Activity 1.**

6. **Record** the results of your tests (positive or negative) in **Table 7-3** and also on the **master chart at the front of the room.**

    Enter a "**+**" if the test was **positive** and a "**−**" if the test was **negative.**

7. **Complete Table 7-3** by entering the results from the **master chart** at the front of the room.

TABLE 7-3

RESULTS OF FOOD EXPERIMENTS

Food	Simple Sugars	Starch	Protein	Lipids
Lettuce				
Hamburger				
Tuna				
Milk				
Refried beans				
Peanut butter				

✔ Comprehension Check

1. List the foods tested that contain **all three** of the following: **carbohydrates, proteins,** and **lipids.**

    It's possible to relate the nutrient content of a food to its original function in plants and animals. Using the information provided below, **explain** why certain nutrients occur in high levels in particular foods.

2. This food comes from the leaf of a plant. Leaf cells perform photosynthesis.

    Which of the six foods is being described?_____

    (Circle one answer.) If photosynthesis is taking place, I would predict a positive indicator test for **simple sugar / lipid / protein.**

3. Which of the foods tested were **positive** for **lipids?**

4. Which of the foods tested use **lipids** as an energy source for **young animals?**

5. **Muscle tissue** is composed of **interwoven protein fibers.** Using this information, **name two foods** made of **muscle** tissue that gave you a **positive** test for **protein.**

6. The hamburger and tuna tested **negative** for starch. **Explain** why foods of this type would **not** contain starch.

Check your answers with your instructor before you continue.

ACTIVITY 3   AN EMBRYO IN A PEANUT SEED

A seed contains a **plant embryo** packaged with its **food supply.** The embryo and stored food are protected by a tough seed coat. A developing embryo has the beginnings of a stem, tiny leaves, and a root. You can see the tiny stem, leaf, and root structures of an **embryonic peanut plant** if you carefully pull the two peanut halves apart.

  The nutrients in seeds are not only used by germinating plant embryos but are also consumed by animals, including humans.

1. In your classroom, you'll find a demonstration of a **separated peanut seed** that can be viewed with a dissecting microscope.

2. Look for the following structures and **label** them on the photo in **Figure 7-1: root, stem, leaves,** and **stored food supply.** Draw a circle around the peanut embryo.

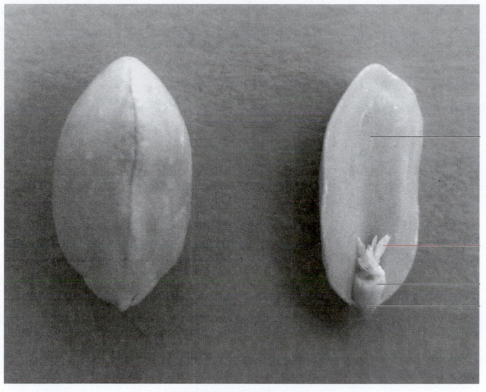

FIGURE 7-1.   Peanut Seed

☑ Comprehension Check

1. Food may be stored for the use of an adult plant or as energy for the development
    of an embryo. The stored food is packaged with an embryo into a structure called
    a **seed.**

    List **three commonly eaten foods** produced by plants or animals that provide
    stored energy for embryos:

    _____   _____   _____

2. Which of the foods tested in Activity 2 use **lipids** as an energy source for a **devel-
    oping embryo?**

3. **(Circle all correct answers.)** The following probably have high lipid levels:

   a. chicken egg   c. sunflower seed   e. coconut

   b. carrot   d. spinach   f. almond

   Explain your answer.

Check your answers with your instructor before you continue.

ACTIVITY 4   DIETARY INTAKE AND GOOD HEALTH

When you eat a meal, you're probably consuming a combination of carbohydrates, proteins, and lipids. Imagine this situation. You've been stuck in classes all morning (just grabbed a bag of peanuts from the candy machine), worked in the afternoon, ran errands, and now you're late for a meeting of your study group. Of course, you're starving! On the way to the library, you stop off at your favorite fast-food restaurant to refuel. You order a double hamburger with cheese, a large order of fries, and a chocolate milk shake.

One person in your study group is taking a nutrition course and has been studying dietary analysis. He was shocked to discover that his normal intake wasn't even close to the guidelines for a healthy diet. As he tells you about this, you wonder whether your diet is any better. Below, you'll find the instructions provided by your friend for calculating the percentage of carbohydrates, protein, and fat in your diet. He also thoughtfully provided you with a copy of the government recommendations for a healthy diet (**Table 7-4**). **Table 7-5** summarizes your meals.

<table>
<tr><td colspan="2">

T A B L E 7 - 4
RECOMMENDED DIETARY GUIDELINES
</td></tr>
</table>

ORGANIC COMPOUND	RECOMMENDED PERCENTAGE OF DAILY CALORIC INTAKE	
Carbohydrates	55–60%	
Protein	10–15%	
Fat	no more than 30%	
1 g carbohydrate contains 4 kcal	1 g protein contains 4 kcal	1 g fat contains 9 kcal

Follow the instructions below to calculate the percentage of carbohydrate, protein, and fat you consumed today.

1. Begin your calculations with the hamburger. **Multiply** the number of grams of carbohydrates in the hamburger by **4** (because there are 4 kcal per gram of carbohydrate).

    Record the answer in Table 7-5. Perform the same calculations for the **protein** and **fat** in the hamburger (using the appropriate number of kilocalories) and record your answers.

2. Calculate the **kilocalories** for the other foods listed in **Table 7-5** and record your answers in the table.

T A B L E 7 - 5
DIETARY CALCULATIONS

FOOD EATEN	CARBO-HYDRATE (g)	CARBO-HYDRATE (kcal)	PROTEIN (g)	PROTEIN (kcal)	FAT (g)	FAT (kcal)
Double beef hamburger with cheese	54		51		60	
Fries (large order)	43		5		20	
Chocolate milk shake	49		9		10	
Bag of peanuts (2 ounces)	10		14		28	
Column Totals						

3. Add the total kilocalories from carbohydrates, proteins, and fats to determine the total kilocalories consumed at this meal.

    Total kilocalories consumed = _____ kcal

4. Calculate the percentage of your diet that came from carbohydrates, protein, and fat.

    % of carbohydrates in diet $= \dfrac{\text{total kcal from carbohydrates}}{\text{total kcal consumed}} \times 100 =$ _____ %

    % of protein in diet $= \dfrac{\text{total kcal from protein}}{\text{total kcal consumed}} \times 100$ $=$ _____ %

    % of fat in diet $= \dfrac{\text{total kcal from fat}}{\text{total kcal consumed}} \times 100$ $=$ _____ %

    100%

5. Compare your calculated percentages to the federal government's recommended values.

    Was your intake today in line with recommended dietary goals? _____

    If not, which nutrients didn't match the guidelines?

6. Place an X in front of any of the following that are good suggestions for improving your day's food intake.

    ___ Bring an apple or an orange for a snack instead of the peanuts.

    ___ Replace the fries with a baked potato.

    ___ Replace the hamburger with pizza.

    ___ Replace the milk shake with low-fat milk or juice.

    ___ Eat a candy bar instead of the peanuts.

7. If your body requires **1500 kcal per day,** how many excess kilocalories did you eat
    in the day described in **Table 7-5?** _____ kcal

    If you had consumed the same amount of excess kilocalories each day for the last
    three weeks, **how many excess kilocalories** did you accumulate? _____ kcal

    Every time you accumulate 3500 excess kilocalories, you gain a pound.

    How many pounds have you gained over the last three weeks? _____ lbs

8. You want to lose this excess weight, but with your schedule, you're too busy to
    exercise regularly. However, if you park in the furthest lot from the building, you
    can get to class in 15 minutes of quick walking. Every time you do this, you'll burn
    150 kcal.

    If you walk quickly **to and from** the parking lot once a day, you can lose the weight
    you gained in only _____ days.

Check your answers with your instructor before you continue.

SELF TEST

1. You add **Biuret solution** to your morning orange juice. The Biuret solution does not change color. What can you **conclude** from this experiment?

2. What do you think would happen if you placed a drop of iodine on your baked potato at dinner?

    On your steak?

    Explain your answer.

3. A class member performed Benedict's test to determine whether potatoes contain simple sugar. To the experimental test tube, she added **ground-up potato, distilled water, 10 drops of glucose solution, and 5 ml of Benedict's solution.** After a few minutes of boiling, the liquid in the test tube turned orange. The student concluded that potato does contain simple sugars. Do you agree with her conclusions? Explain your answer.

4. List two changes you could make in your typical daily meals to come closer to the recommended dietary guidelines. Explain why each change would help you achieve a healthier diet.

Factors That Affect Enzyme Activity

Objectives

After completing this exercise, you should be able to:

- discuss the basics of enzyme function in cells
- explain the relationship between the three-dimensional structure of proteins and enzyme function
- describe the effects of various environmental conditions on protein denaturation
- explain the activity of digestive enzymes in food vacuoles

CONTENT FOCUS

Within living systems, chemical reactions require specific **enzymes** to assist and speed the rate of the reactions (these "helper" molecules are known as **catalysts**). Most enzymes are **proteins** and they are very specific, each working with only one or very few chemical compounds. In addition, different enzymes work best under different environmental conditions. One way enzymes help chemical reactions occur is by reducing the energy input needed for the reaction to take place.

In order for a protein to function correctly, it must be folded into a specific three-dimensional shape. In enzyme molecules, folding creates a region called the **active site,** where molecules can bind to the enzyme and a reaction can take place. Each type of enzyme works for only particular reactions, because its active sites will accept only molecules (called **substrates**) that have matching shapes (see **Figure 8-1**). Note that the enzyme is not permanently altered by the reaction and can perform the same function multiple times.

If environmental factors (such as pH, temperature, or salinity) lead to changes in the specific three-dimensional shape of the enzyme's active site, the enzyme may not function correctly. When an enzyme's folding is altered, the enzyme is said to be **denatured.**

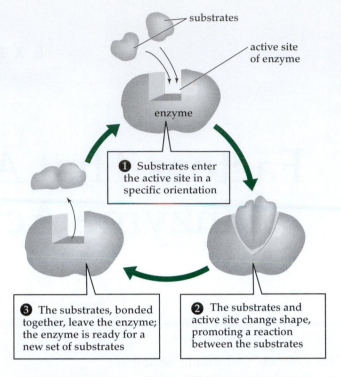

FIGURE 8-1.   Enzyme/Substrate Interaction

The following activities will demonstrate how enzymes work and the effect of some environmental conditions on protein structure.

ACTIVITY 1   DEMONSTRATION OF ENZYME ACTIVITY

Amylase is a digestive enzyme that **hydrolyzes starch** (a polysaccharide). Hydrolysis is a chemical reaction in which larger molecules are broken into smaller components. In humans, amylase is present in the mouth and small intestine.

To demonstrate the action of amylase on starch molecules, two indicator tests will be used.

Iodine is a chemical indicator that changes color (turns dark blue or black) in the presence of **starch.** In the presence of monosaccharides or disaccharides, iodine retains its normal reddish-brown color.

Benedict's solution is a chemical indicator for the presence of **simple sugars** (monosaccharides and some disaccharides). In the presence of simple sugars, Benedict's solution changes color from turquoise blue to one of the following colors: green, yellow, orange, or red. Green represents the smallest amount of simple sugar and red the greatest. Benedict's test differs from the iodine test in that the reaction takes place only when the solution is heated.

1. Work in groups. Get the following supplies: **four large test tubes, two glass stirring rods, a test tube rack, a test tube holder, two graduated 10-ml pipettes with manual dispenser, a hot plate, a dropper bottle of iodine solution, a dropper bottle of Benedict's solution, a stock bottle of starch solution, a stock bottle of amylase solution, and a large beaker.**

2. Place the **large beaker,** about **one-third filled** with tap water, on the **hot plate.**

    Set the temperature control to high until the water boils.

3. **Label the test tubes** as follows: **B-1, B-2, I-1,** and **I-2.**

    Make sure the bottle of starch solution is **well mixed.**

    Using a **10-ml graduated pipette,** transfer **15 ml of starch solution** from the stock container into **each** of the large test tubes.

4. To the test tube labeled **B-1,** add **two droppers full of Benedict's solution.**

    What color is the liquid in **tube B-1?** _____

5. The chemical indicator in Benedict's solution **will react only with simple sugars when it is heated.**

    Using a **test tube holder, carefully** place the **tube B-1** into the boiling water for **two minutes.**

Caution!

Hot glass looks exactly like cold glass! Use the test tube holder to remove test tubes from the boiling water!

Don't leave the test tube holder clamped to the test tube while in the boiling water. Hot metal looks exactly like cold metal!

---

6. **Using the test tube holder, remove** the test tube from the boiling water and **observe** the color of the Benedict's solution.

    Don't dispose of tube B-1.

    Was your Benedict's test result for tube **B-1 positive** or **negative (+/ −)?**

7. To the test tube labeled **I-1,** add two droppers full of iodine solution.

    Don't add indicator solution to test tube **I-2.**

8. Observe the color of the indicator solution in **tube I-1**.

    Was your iodine test result **positive or negative (+/ −)?**

    Don't dispose of tube I-1.

9. Using a **clean 10-ml graduated pipette,** add **15 ml of amylase solution** to **tubes B-2 and I-2.**

    Record the time when you add the amylase to the starch solution: _____

    Place a **glass stirring rod** into each of the tubes **B-2** and **I-2.**

    Set tubes B-2 and I-2 aside for **30 minutes.**

    Every five minutes, stir the contents of each tube with the rod.

10. After the 30-minute interval, add **two droppers full of Benedict's solution to tube B-2.**

    Add two droppers full of iodine solution to tube I-2.

    Shake the test tubes to mix the contents.

11. Observe the color of the indicator solution in **tube I-2.**

    Is your iodine test result **positive or negative (+/ −)?**

12. If the **iodine test is negative in tube I-2,** what would this tell you about the digestion of starch by amylase?

13. Imagine that you get a **positive result for the iodine test in tube I-2.** You form the **following hypothesis** about the reason for the positive test result:

   The amylase enzyme was inactive and therefore didn't digest any starch.

   Suggest a method by which this hypothesis could be tested.

14. **Using the test tube holder,** place **tube B-2** in the boiling water for **two minutes.**

   Remove the test tube from the boiling water and **observe** the color of the Benedict's solution.

   Was your Benedict's test result **positive or negative (+/ −)?**

15. Do the test results **support your hypothesis? Explain** your answer.

✓ Comprehension Check

   1. If your Benedict's test result was positive, where did the sugar come from? **Explain** your answer.

2. What was the purpose of the indicator tests conducted on tubes **I-1** and **B-1**?

3. Why were tubes **I-1** and **B-1** saved until the end of the experiment?

Check your answers with your instructor before you continue.

ACTIVITY 2   ENZYME ACTIVITY IN FOOD VACUOLES

Paramecium is a small one-celled organism found in freshwater. By means of rapid swimming motions, it funnels water containing bits of organic matter, such as yeast and bacteria, into a groove on its surface. From this groove, the materials enter the *Paramecium* through a "mouth" and are taken up by small organelles called **food vacuoles.**

   The organic matter constitutes the *Paramecium*'s food. Enzymes are released into the vacuoles and the food within them is digested.

1. Work in groups. Get the following supplies: **slides, coverslips, a compound microscope, a bottle of methyl cellulose (the bottle may also be labeled Protoslo®), toothpicks, a *Paramecium* culture, and a solution of yeast stained with Congo red.**

2. Make a wet mount using the *Paramecium* culture and add **one drop** of **red-stained yeast.**

    Add a **small** drop of the methyl cellulose to your slide, **gently** stir with a toothpick, and add a coverslip.

3. Using the **scanning lens** of the microscope, locate one or more organisms.

    Switch to a higher power and locate the food vacuoles within the cell (refer to **Figure 8-2**).

    What is the initial color of the food vacuoles? _____

    Observe the food vacuoles in various individuals every few minutes for **15 minutes.**

    Describe the **color changes** that occur within the vacuoles.

4. Congo red is an **indicator chemical** that turns **blue** under **acidic** conditions and **red** under **basic** conditions.

    What does this tell you about the pH inside a food vacuole?

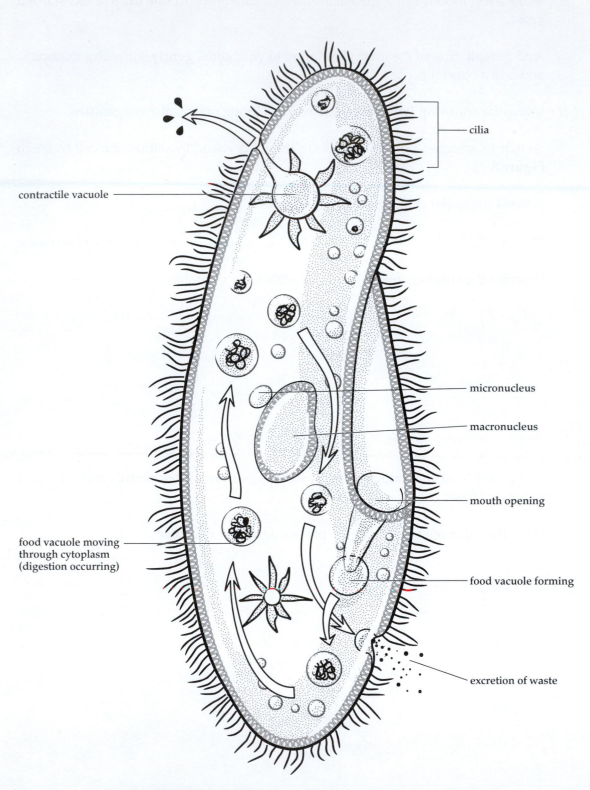

cilia

contractile vacuole

micronucleus

macronucleus

mouth opening

food vacuole moving
through cytoplasm
(digestion occurring)

food vacuole forming

excretion of waste

FIGURE 8-2. *Paramecium* Food Vacuole Formation

✓ Comprehension Check

1. **Pepsin** is a human digestive enzyme that functions in the acidic environment of the stomach. If pepsin were present in *Paramecium*, do you think this enzyme could be active within a food vacuole? **Explain** your answer.

2. If a **base (alkaline solution)** was added to the yeast suspension before the *Paramecia* were fed, how might this affect the **color change** within the food vacuoles? **Explain** your answer.

3. If you ground up a "Tums," an antacid tablet, to a fine powder and added it to the yeast suspension before the *Paramecia* were fed, how might this affect the **efficiency of digestion within the food vacuoles?**

Check your answers with your instructor before you continue.

ACTIVITY 3
EFFECT OF ENVIRONMENTAL CONDITIONS ON PROTEIN STRUCTURE

During this laboratory exercise, you'll be examining the effect of various environmental factors (temperature, salinity, and pH) on the folded structure of proteins. You'll use chicken eggs because they are among the few cells that are large enough to be observed without special equipment.

1. Work in groups. Get the following supplies: **five chicken eggs; five bowls; stock bottles of the following solutions: pH 11, pH 3, and 25% saline; a large beaker; clear plastic wrap; and a hot plate.**

2. Pour about **400 ml of tap water** into the large beaker and set the hot plate on **high.** Heat the water to boiling.

3. **Form hypotheses** about the effect of each set of environmental conditions **(pH 11, pH 3, 25% saline, boiling, and room temperature)** on the yolk and the white of an egg.

   For example: no change will occur in the yolk or the white, the proteins in the egg yolk will denature, the proteins in the egg white will denature, the egg white will denature but not the yolk, and so on.

   Record your hypotheses in **Table 8-1.**

TABLE 8-1 RESULTS OF EGG EXPERIMENTS		
ENVIRONMENTAL CONDITIONS	HYPOTHESES	OBSERVATIONS OF THE EGG WHITE AND YOLK
pH 11		
pH 3		
25% saline solution		
Boiling water		
Room temperature		

4. Label **three** of the bowls as follows: **pH 11, pH 3,** and **25% saline.**

    Carefully, without breaking the yolk, crack an egg into each of the three labeled bowls.

Caution!

Be careful not to spill the pH solutions on your skin. Rinse your hands or eyes thoroughly with water if contact occurs.

---

5. To the bowl labeled **pH 11,** add **100 ml of pH 11** solution.

    To the bowl labeled **pH 3,** add **100 ml of pH 3** solution.

    To the bowl labeled **25% saline,** add **100 ml of 25% saline** solution.

    Set the bowls aside undisturbed for **30 minutes.**

6. **Carefully, without breaking the yolk,** crack an egg into each of the two remaining bowls.

    To minimize odors, loosely cover the **pH 11** bowl with a sheet of clear plastic wrap.

    Set one of the bowls aside undisturbed for **30 minutes.**

7. Gently tip the egg out of the other bowl and into the beaker of boiling water.

    Let the water return to a full boil and boil the egg for **five minutes.**

    Turn the hot plate off and let the water cool to room temperature.

8. At the end of the 30-minute period, observe and record your observations for the yolk and white of each egg in **Table 8-1.**

Note:

Don't dispose of any of the eggs until you have answered the remaining questions.

9. Did any of the environmental conditions damage **(denature)** the proteins in the egg? **Explain** your answer.

10. Test the white part of the **"room temperature"** egg with pH paper.

    What is the approximate pH of the egg white? _____

    Is there a relationship between the pH results from the **room temperature** egg and your observations of the eggs exposed to **pH 3** and **pH 11? Explain** your answer.

11. Was there a **control** in this experiment? If so, what was the control?

12.   Why is a control desirable in an experiment of this type?

Comprehension Check

1.   Stomach conditions are quite acidic. Do cells of the stomach lining need protection from these acidic conditions? Why or why not?

2.   Acid rain in the Northeastern United States can have a pH as low as 4.5 in some areas. How might this explain decreased survival among fish and amphibian eggs in these regions?

3. With reference to question 2 above, why might the effect on bird eggs be **less serious** than that experienced by fish and amphibian eggs?

 Check your answers with your instructor before you continue.

SELF TEST

1. Based on the results of your experiments, why is it considered dangerous to have an abnormally high fever?

2. Body fluids contain buffers whose function is to minimize changes in pH levels. Why is a buffering system beneficial for optimal enzyme function?

The following are the results of an experiment that examined the effect of pH on the activity of an enzyme isolated from the human digestive tract.

Tube #	pH	Elapsed Time (minutes)											
		2	4	6	8	10	12	14	16	18	20	22	24
1	2.0	−	−	−	−	−	−	−	−	−	−	−	−
2	3.0	−	−	−	−	−	−	−	−	−	−	−	−
3	3.3	−	−	−	+	+	+	+	+	+	+	+	+
4	3.5	+	+	+	+	+	+	+	+	+	+	+	+
5	3.7	−	−	−	−	−	−	+	+	+	+	+	+
6	4.0	−	−	−	−	−	−	−	−	−	−	−	−

3. The optimal pH for this enzyme is _____.

4. In a few sentences, summarize the experimental results for **pH 3.7.**

5. Would you characterize this enzyme as **general or specific** in regard to its pH requirements? **Explain** your answer.

6. Enzymes regulate the production of the pigment melanin in the fur of Siamese cats and Himalayan rabbits. At normal body temperatures, the rabbits produce white fur, but their paws, ear tips, and noses have black fur. On the basis of your understanding of the effect of environmental conditions on enzyme activity, suggest one possible explanation for the difference in fur color on various parts of the body.

Introduction to Molecular Genetics

Objectives

After completing this exercise, you should be able to:

- extract DNA from a tissue sample
- demonstrate an understanding of DNA structure and the base-pairing rule
- explain the roles of the following in protein synthesis: DNA, mRNA, tRNA, and ribosomes
- complete the processes of transcription and translation from a strand of DNA and determine the sequence of amino acids in the resulting polypeptide chain
- explain how changes in the DNA code may affect protein synthesis

CONTENT FOCUS

Chromosomes are divided into sections called **genes** that are the basis of inheritance. Genes contain the coded instructions your body uses to assemble the hundreds of different types of proteins that make you a unique individual. Genes are composed of a molecule known as **deoxyribonucleic acid (DNA).** In this exercise, you'll take a closer look at the DNA molecule and the role it plays in all the cells of your body.

The traits you inherited through your genes (such as hair color, blood type, presence of sickle-cell anemia, or ability to produce the hormone insulin) are all controlled by the production of proteins. The process of protein synthesis is divided into two general steps. **Transcription,** which occurs in the nucleus of eukaryotic cells, forms strands of messenger RNA (mRNA). An mRNA strand is a copy of the genetic instructions to make one polypeptide chain.

The bonding of amino acids into proteins by ribosomes is called **translation.** This process occurs in the cytoplasm. Amino acids are picked up in the cytoplasm by molecules of transfer RNA (tRNA), which carry them to the ribosomes for assembly.

ACTIVITY 1 REMOVING DNA FROM CELLS

> ### Note:
> **Read these directions COMPLETELY before proceeding.**

1. Work in groups. Get the following supplies: **one piece of calf thymus (5 grams), one graduated 10-ml pipette with manual dispenser, one 50-ml test tube, a test tube rack, a small bowl, and a pair of scissors.**

2. On your laboratory table, you'll find a **blender, a piece of cheesecloth, a 500-ml graduated cylinder, a container of 6% saline solution, rubber bands, and a 600-ml beaker.**

   Using **scissors, mince** the piece of thymus **as much as possible** and place the pieces in the bowl.

   Add **100 ml of 6% saline solution** to the bowl.

> ### Note:
> **All groups at your table will be using the blender at the same time, so you MUST coordinate your activities!**

3. Coordinating with your partner groups, pour the saline containing the **minced thymus** from **each of the groups** at your laboratory table into the blender.

4. **Put the lid on the blender.** Blend the thymus mixture **on a low setting** for several minutes until no large pieces remain.

5. **Remove the lid** from the blender.

   Assemble **eight thicknesses of cheesecloth** (four pieces from the package, which is manufactured in double thickness). Place the pieces of cheesecloth over the top of the blender and **secure** the cloth **tightly** with the rubber band.

6. Making sure the cheesecloth is **tightly** fastened on the top of the blender, **pour the contents (filtered extract)** into the **600-ml beaker.**

> ### Note:
> **At this point, you'll go back into your ORIGINAL groups and complete the experiment.**

7. Using a **graduated 10-ml pipette with manual dispenser,** remove **10 ml of filtered extract** from the beaker and place the extract into a **50-ml test tube.** Place the test tube in the rack.

    Place your **used pipettes** in the waste container.

8. Get **one graduated 2-ml pipette with manual dispenser, one graduated 10-ml pipette, and a thin glass rod.**

    In your supply area, you'll find a container of **10% SDS solution** (sodium dodecyl sulfate) and **an ice bucket containing 15-ml screw-cap tubes of 95% ethanol.**

9. To the **test tube containing the filtered thymus** extract, add **1 ml of SDS solution.** Tap the test tube several times to **gently** mix the contents.

Note:

Read the following directions COMPLETELY before proceeding.

---

10. Select a **tube of ethanol** from the ice bucket. Using a **clean,** graduated 10-ml pipette, add **10 ml of ice-cold 95% ethanol** to the test tube.

    With the test tube in the rack, place the filled pipette with its tip against the **inside wall** of the test tube. **SLOWLY** allow the ethanol to dribble down the inside of the tube, as **demonstrated in Figure 9-1. Don't shake the test tube during this procedure.**

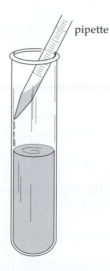

pipette

FIGURE 9-1. Method for Adding Ethanol to Filtered Thymus Extract

11. The ethanol is lighter than the contents of the tube. When added according to directions, the ethanol will form a **clear layer ABOVE** the filtered thymus extract (see **Figure 9-2**).

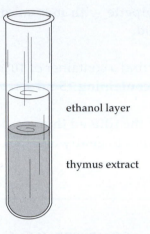

ethanol layer

thymus extract

FIGURE 9-2.   Ethanol Layered Above Filtered Thymus Extract

12. Let the DNA sample remain undisturbed for at least 10 minutes. The DNA will gradually separate from the thymus mixture and rise into the ethanol layer.

Describe the appearance of the DNA.

13. To remove the accumulated DNA from the test tube, follow the directions for **DNA spooling** below:

   a. **Gently insert** the glass rod through **the ethanol layer** into the **accumulated DNA.**

   b. **Carefully twirl** the rod between your fingers, winding the DNA onto the rod, imitating thread on a spool.

   c. **Slowly remove the rod** from the test tube.

✔ Comprehension Check

1. The thymus is an organ made up of many different types of cells. What percentage of thymus cells contain DNA?

   a. 0%   d. 75%

   b. 25%   e. 100%

   c. 50%

   Explain your answer.

2. If you removed DNA from a sample of human thymus, how many separate DNA molecules would be removed from **each cell?** _____

3. **(Circle one answer.)** If you repeated the same experiment with an equal number of kidney cells, the amount of DNA collected would **increase / decrease / stay the same.**

   Explain your answer.

4. **(Circle one answer.)** If you repeated the same experiment with an equal number of sperm cells, the amount of DNA collected would **increase / decrease / stay the same.**

   Explain your answer.

5. **Challenge Question! (Circle one answer.)** If you repeated the same experiment with an equal number of red blood cells, the amount of DNA collected would **increase / decrease / stay the same.**

    Explain your answer.

Check your answers with your instructor before you continue.

ACTIVITY 2   THE BASICS OF DNA STRUCTURE

The extraction of DNA that you just performed showed that DNA is present in cells. It does not, however, give you much information about its actual structure.

DNA molecules are composed of small building blocks called **nucleotides.**

1. Each DNA nucleotide is composed of three smaller molecules hooked together:

    one five-carbon sugar (deoxyribose)

    one phosphate

    one nitrogen base

2. Four different types of nucleotides are needed to build a DNA molecule.

    Each of these four nucleotides has a **different nitrogen base: adenine, guanine, cytosine, or thymine.**

3. DNA has a structure similar to a ladder. The two sides of the ladder are composed of **alternating sugar and phosphate molecules.**

4. Each sugar molecule is attached to **one nitrogen base.** The two strands of DNA are attached by bonds between the nitrogen bases on each side of the ladder.

    Each nucleotide base only bonds with **one specific partner.** The combination of two bases is called a **base pair.**

Adenine always bonds with thymine. A-T

Guanine always bonds with cytosine. G-C

5. Fill in the blanks on the incomplete DNA molecule in **Figure 9-3.**

Use the symbol **"P"** for **phosphate, "S"** for **sugar,** and **"A," "T," "C,"** or **"G"** for the **appropriate nitrogen bases.**

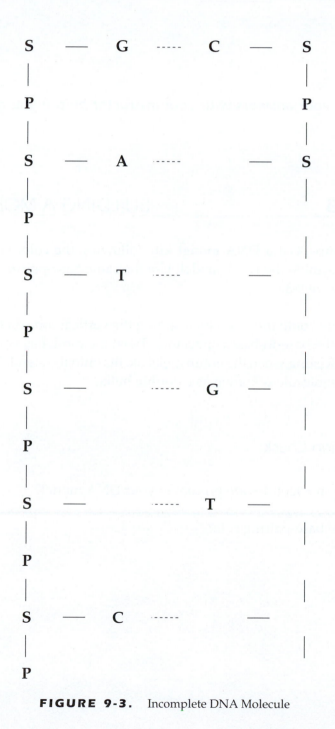

FIGURE 9-3. Incomplete DNA Molecule

✔ Comprehension Check

1. On **Figure 9-3**, draw a box around **one complete nucleotide.**

2. **How many nucleotides** are shown in the DNA molecule in **Figure 9-3?** _____

3. How many **different** types of nucleotides were used to construct the DNA molecule in **Figure 9-3?** _____

Check your answers with your instructor before you continue.

ACTIVITY 3   BUILDING A MODEL OF DNA

1. Work in groups. Get a **DNA model kit.** Following the color-coded instructions in the kit, assemble the DNA model. Use the **same** base-pairing sequence as that shown in **Figure 9-3.**

2. Place the model onto the stand by inserting the vertical tube on the stand through the holes in the base-to-base connectors. Twist the model as you slide it onto the stand. Stop twisting when the entire molecule fits onto the stand. The twisted shape of the DNA molecule is known as a **double helix.**

✔ Comprehension Check

1. How many nucleotides are present in your DNA model? _____

2. What is the base-pairing rule?

3. Assume that the model you just built is an exact representation of **your** DNA code.

    a. Would you use the same bases to construct your lab partner's DNA? _____

    b. Would you assemble the bases in the same order to make a model of your lab partner's DNA? _____

    Explain your answers.

Check your answers with your instructor before you continue.

Genes have the instructions to make polypeptide chains. These instructions are part of the genetic code. Polypeptide chains are the structural units of proteins. A polypeptide chain is made of many amino acids hooked together.

The key to the genetic code is the sequence of nitrogen bases along **one side** of the DNA molecule. To construct a protein, you must know the **order of the bases.** The code is written in **three-letter "words."** Each of these words (called **triplets**) tells the cell which amino acid should come next when building a protein.

For proteins to function properly, the amino acids must be assembled in the correct order.

3. How many **triplets** are present along one side of your DNA model? _____

4. How many **amino acids** will be present in the protein made from your model? _____

ACTIVITY 4   STEPS OF PROTEIN SYNTHESIS

When a particular protein is needed by the body, regions of the double helix unwind so that a cell gains access to the genes that contain the coded information to make that protein. Protein synthesis has two steps: **transcription** takes place in the nucleus and **translation** occurs in the cytoplasm. Both steps require molecules of **RNA (ribonucleic acid).**

Although the **nucleus** contains the **instructions** for protein synthesis, the **machinery to make proteins** is located in the **cytoplasm.** The coded information is **transferred** from the nucleus to the cytoplasm during **transcription.**

Genes begin with an **initiator (start) codon** (AUG) that also codes for the amino acid methionine. In addition, the end of each mRNA sequence has a **terminator (stop) codon** that signals the ribosome that it is at the end of the amino acid sequence.

For example, consider the mRNA transcription of the following DNA sequence:

DNA: T A C   A G A T A A C C C G C G   A C T

mRNA: **A U G   U C U A U U G G G C G C   U G A**

| **start codon** | coding region | **stop codon** |

Transcription

1. During transcription, DNA bases are **copied** to form a single strand of RNA, called **messenger RNA (mRNA).** As with the DNA, mRNA is divided into coded three-letter words. In mRNA, these words are called **codons.**

2. The **base-pairing rule** is used to form messenger RNA **with one exception.** RNA molecules **don't** have the nitrogen base **thymine.** They have **uracil** instead.

Base-pairing to form mRNA:

DNA BASE	MRNA BASES
C	G
G	C
T	A
A	**U**

3. The coded information to make a protein appears along **one side** of the double helix. Practice transcription by filling in the correct messenger RNA codons in **Table 9-1.**

TABLE 9-1 TRANSCRIPTION OF mRNA								
DNA TRIPLETS	CGC	ATA	GAC	TTT	CTT	GAT	TAG	CAT	AAA
mRNA CODONS									

> ### Note:
> To make it easier for you to practice transcription and translation, the start and stop codons have been removed from all the DNA sequences and only the coding regions of the DNA are shown.

4. How many **amino acids** would be in this protein? _____

5. Which of the five types of nitrogen bases is **not** found in mRNA? _____

Translation

1. A cell needs **amino acids** to construct proteins. The amino acids are carried to the ribosomes by another type of RNA molecule, called **transfer RNA (tRNA).**

   A tRNA molecule has **two functional ends.**

   One end picks up amino acids in the cytoplasm (see **Figure 9-4**).

2. The other end is called the **anticodon.** It contains **three nitrogen bases** that can form a base pair with a **matching codon** in the messenger RNA.

3. Each type of tRNA can carry **only one type of amino acid.**

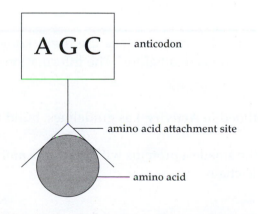

FIGURE 9-4. Structure of a Transfer RNA Molecule

There are enough different types of transfer RNA molecules to carry all the different types of amino acids needed to make your body's proteins.

4. **Where do the transfer RNA molecules take the amino acids?** They take them to **ribosomes**, organelles in the cytoplasm where proteins are manufactured. Ribosomes are made of proteins and a third type of RNA called **ribosomal RNA.**

5. **Ribosomes read messenger RNA codons and accept amino acids brought by transfer RNA molecules.** Ribosomes hook amino acids together in the **order specified by the messenger RNA** codons to construct the polypeptide chain.

Summary of Protein Synthesis

■ DNA contains the instructions to make polypeptide chains. A region of the DNA double helix unwinds. Coded instructions to make a protein are exposed.

■ This information is carried to the cytoplasm by messenger RNA molecules (transcription).

■ Amino acids in the cytoplasm are used to build polypeptides.

■ Transfer RNA molecules pick up the amino acids and transport them to ribosomes, the locations where proteins are made.

■ Ribosomes bond amino acids together according to the instructions in the genetic code.

ACTIVITY 5   BUILDING A REAL PROTEIN

Imagine the following situation: you're about to give birth to a baby. The brain produces the hormone **oxytocin** (a small protein), which causes uterine muscles to contract for childbirth. After birth, this same hormone causes muscles in the mammary glands to contract, releasing milk for nursing the baby.

Suppose this is your first baby. How does the brain know how to manufacture oxytocin if it has never been needed before? The information is stored in your DNA "reference" library.

1. Using the steps outlined in **Activity 4** as guidelines, build the protein **oxytocin.**

    Oxytocin is one of the smallest proteins with only nine amino acids connected into a single polypeptide chain.

2. **Transcribe** the DNA triplets that code for oxytocin into **messenger RNA codons** and add them to the appropriate spaces in **Table 9-2.**

    Reminder: The start and stop triplets have been removed from this DNA strand.

T A B L E 9 - 2 TRANSCRIPTION OF OXYTOCIN									
DNA TRIPLETS	ACG	ATG	TAT	GTT	TTG	ACG	GGA	GAC	CCC
mRNA CODONS									

3. The messenger RNA must detach from the DNA and leave the nucleus for **translation** to take place.

    Cut out the strip of mRNA and move it to the ribosome in Figure 9-5. Notice that the ribosome has two parts, separated by a groove. The mRNA should be placed along the groove between the upper and the lower sections of the ribosome, over the words "mRNA binding site."

4. Each transfer RNA molecule in **Figure 9-6** is ready to deliver an amino acid to the ribosome.

    Where should each transfer RNA molecule deliver its amino acid?

    Remove Figure 9-6 from your lab manual and cut out the transfer RNA molecules.

    Place each tRNA molecule under the correct mRNA codon.

> ### Hint:
> You'll be able to match each tRNA with its codon by following the base-pairing rule.

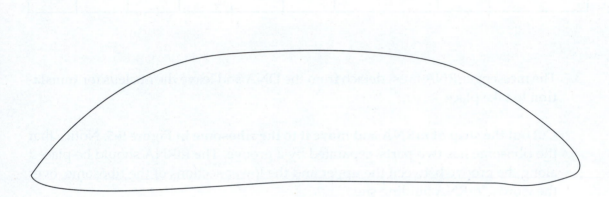

mRNA binding site

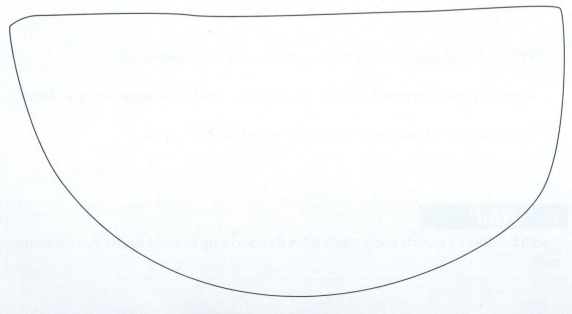

FIGURE 9-5. Ribosome Subunits

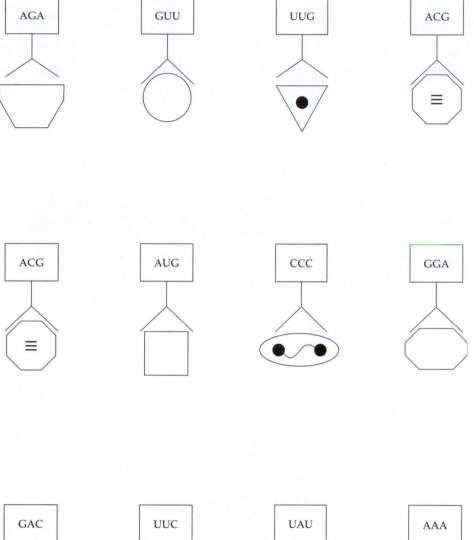

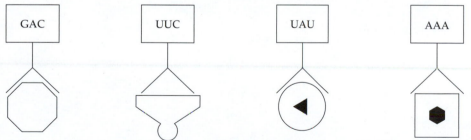

FIGURE 9-6. Molecules of tRNA

To determine the **sequence of amino acids in the protein oxytocin,** match the **shape** of each amino acid **symbol** (attached to the bottom of the tRNA molecules in **Figure 9-6**) to the **amino acid names** in **Table 9-3.** To do this, read the tRNA sequence from **left to right.**

Fill in the blanks below with the names of the amino acids.

Oxytocin amino acid sequence:

Amino acid #1 _____

Amino acid #2 _____

Amino acid #3 _____

Amino acid #4 _____

Amino acid #5 _____

Amino acid #6 _____

Amino acid #7 _____

Amino acid #8 _____

Amino acid #9 _____

Check your oxytocin molecule with your instructor before you continue.

T A B L E   9 - 3
KEY TO IDENTIFYING AMINO ACIDS

SHAPE	NAME OF AMINO ACID
	histidine
	glutamine
	asparagine
	cysteine
	tyrosine
	glycine
	proline
	leucine
	alanine
	isoleucine
	serine

✓ Comprehension Check

1. If the second amino acid in your polypeptide chain were changed to **valine,** would the protein still be oxytocin? _____ **Explain** your answer.

2. **(Circle one answer.)** The DNA triplet A A A would be transcribed into the mRNA codon **T T T / U U U.**

3. Arrange the following steps of protein synthesis in the correct order.

   _____ tRNA molecules pick up amino acids

   _____ mRNA transcribed

   _____ DNA double helix unwinds

   _____ mRNA binds to ribosome

   _____ Ribosome bonds amino acids together

   _____ tRNA anticodon links with mRNA codon

   _____ mRNA leaves nucleus

   _____ Polypeptide chain completed

4. If you include the start and stop codons that were removed from the mRNA sequence for oxytocin, how many DNA triplets were in the **original gene** that coded for oxytocin? _____

   Based on your answer to the question above, how many **additional nucleotide bases** would have been present in the **original transcribed mRNA** strand for oxytocin? _____

Check your answers with your instructor before you continue.

SELF TEST

Matching Answers can be used **more than once.**

a. DNA   c. both DNA and RNA

b. RNA   d. neither DNA nor RNA

1. _____ Present in chromosomes.

2. _____ Contains the base adenine (A).

3. _____ Structure is a double helix.

4. _____ Contains the base thymine (T).

5. _____ Is made of amino acids linked together.

6. _____ Part of the structure of ribosomes.

7. _____ Made of nucleotides linked together.

8. _____ Contains the base uracil (U).

9. _____ Contains the sugar deoxyribose.

10. _____ Contains phosphate.

11. A friend of yours has volunteered for a study on a new type of gene therapy. He tells you that the researchers intend to examine his circulating red blood cells to determine whether the gene was successfully inserted into a chromosome. Has your friend misunderstood the planned procedures? **Explain your reasoning.**

12. In your own words, explain the relationship among the following terms: **gene, DNA triplet, codons, amino acids,** and **protein.**

We have a "mini-gene" with the sequence **T A T C C T G A T T C A A A A G T T.** Given this information, answer the following questions:

13. Assuming that there are no start or stop triplets present, how many amino acids would be in the protein encoded by this gene? _____

14. The last codon in the mRNA will be _____.

15. The anticodon of the tRNA which carries the **LAST** amino acid will be _____.

16. Using **Table 9-4,** list the **FIRST FOUR** amino acids of this protein in their **correct sequence.**

    Amino acid #1   _____

    Amino acid #2   _____

    Amino acid #3   _____

    Amino acid #4   _____

17. If a mutation changed the DNA code of the second triplet from **C C T to C C C,** would that change the structure of the protein made using this code? **Explain** your answer.

TABLE 9-4
MESSENGER RNA CODONS AND THEIR CORRESPONDING AMINO ACIDS

1ST LETTER	2ND LETTER	3RD LETTER	AMINO ACID	1ST LETTER	2ND LETTER	3RD LETTER	AMINO ACID
A	A	A	lysine	U	A	A	terminator 1
		C	asparagine			C	tyrosine
		U	asparagine			U	tyrosine
		G	lysine			G	terminator 2
	C	A	threonine		C	A	serine
		C	threonine			C	serine
		U	threonine			U	serine
		G	threonine			G	serine
	G	A	arginine		G	A	terminator 3
		C	serine			C	cysteine
		U	serine			U	cysteine
		G	arginine			G	tryptophan
	U	A	isoleucine		U	A	leucine
		C	isoleucine			C	phenylalanine
		U	isoleucine			U	leucine
		G	methionine/start			G	leucine
C	A	A	glutamine	G	A	A	glutamic acid
		C	histidine			C	aspartic acid
		U	histidine			U	aspartic acid
		G	glutamine			G	glutamic acid
	C	A	proline		C	A	alanine
		C	proline			C	alanine
		U	proline			U	alanine
		G	proline			G	alanine
	G	A	arginine		G	A	glycine
		C	arginine			C	glycine
		U	arginine			U	glycine
		G	arginine			G	glycine
	U	A	leucine		U	A	valine
		C	leucine			C	valine
		U	leucine			U	valine
		G	leucine			G	valine

Mitosis and Asexual Reproduction

Objectives

After completing this exercise, you should be able to:

- name and describe the stages of mitosis
- identify the stages of plant and animal mitosis as viewed through the compound microscope
- discuss the relationship between the number of cells in each of the stages of mitosis and the length of the various stages
- explain the relationship between mitosis and the processes of regeneration and asexual reproduction
- apply your knowledge of asexual reproduction to cloning and other medical technologies

CONTENT FOCUS

Most of us are aware that the outer layer of our skin is subject to constant wear. To keep this protective layer intact, skin cells must be replaced throughout our lives. Cells of the epidermis, for example, are replaced every 25 to 45 days. A similar situation exists in some parts of the body, whereas in other tissues and organs, the rate of cell replacement is much slower. Brain cells and heart muscle cells are among the slowest to divide; red blood cells, fingernails, the intestinal lining, and other cells divide rapidly throughout our lives.

Growth is another activity for which additional body cells are required. A **zygote** (fertilized egg) begins life as a single cell that multiplies into many cells as the embryo develops. During childhood, cell multiplication provides the many cells needed for growing into an adult.

The type of **cell division** that makes growth and replacement possible is called **mitosis.** All body cells produced by mitosis must contain **the same genetic information** as all other body cells—the information that makes you a unique individual.

ACTIVITY 1   HOW MITOSIS WORKS

A cell is genetically "programmed" to carry out its function in the body. These cellular instructions are included within structures referred to as **chromosomes.** A full set of these genetic instructions is necessary if a cell is to function normally. Therefore, when new body cells are produced by **mitosis,** each has a complete set of chromosomes.

The normal number of chromosomes in a cell is referred to as the **diploid number (abbreviated 2n).** Each type of organism has a characteristic number of chromosomes. The diploid number of chromosomes in human body cells, for example, is 46. The diploid number in one species of pine tree is 24, in crayfish it's 200, and in fruit flies it's only 4.

Assume that the diploid number of chromosomes in the cell below equals 16.

If this cell divides, how many chromosomes must be present in each new cell?

Enter this number in each of the blank circles below.

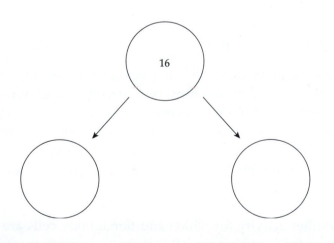

A cell's normal activity pattern is called the **cell cycle.** Cell division is only a small percentage of the cell cycle and includes three main components: DNA replication (a duplication of the cell's chromosomes that occurs during interphase), mitosis, and cyto-kinesis (the physical separation of the "parent" cell into two daughter cells).

To make it easier to understand, the process of mitosis is divided into four stages plus cytokinesis. At the completion of cell division, two new cells are formed. Each cell will have a complete set of genetic instructions.

To complete cell division successfully, **dividing cells must solve several problems:**

■ **Each new cell needs a full set of chromosomes.**

Accomplished by: Duplicating chromosomes (an exact copy is produced).

To avoid confusion, these duplicate chromosomes are referred to as **sister chromatids.**

A cell preparing to divide will contain two complete sets of sister chromatids (one set for delivery to each of the two new cells).

■ **Sister chromatids must be attached together.**

Accomplished by: A structure called a **centromere** fastens the duplicates together.

Attachment makes it easier to keep track of sister chromatids for sorting into two groups (one set for each new cell).

■ **Pieces of chromosomes shouldn't get broken off or lost.**

Accomplished by: The long, threadlike chromosomes coil and fold into compact structures. Condensed chromosomes are thick enough to be visible with the compound microscope.

■ **The pairs of sister chromatids must be pulled apart and delivered to each new cell.**

Accomplished by: A structure called the **mitotic spindle** hooks on to each pair of sister chromatids. **Spindle fibers** pull the chromatids apart and deliver one chromatid from each pair to each new cell.

■ **When the sorting and delivery process is complete, the two cells must be separated into two daughter cells, each with its own set of chromosomes.**

Accomplished by: A process called **cytokinesis** separates the cytoplasm into two halves (**"cyto"** means *cell* and **"kinesis"** means *cutting*).

During cytokinesis in animal cells, the cell membrane pinches into a groove called the **cleavage furrow.**

ACTIVITY 2   RECOGNIZING THE STAGES OF MITOSIS

Use the **Decision Tree in Figure 10-1** to **identify the stages of the cell cycle** in **Figure 10-2.**

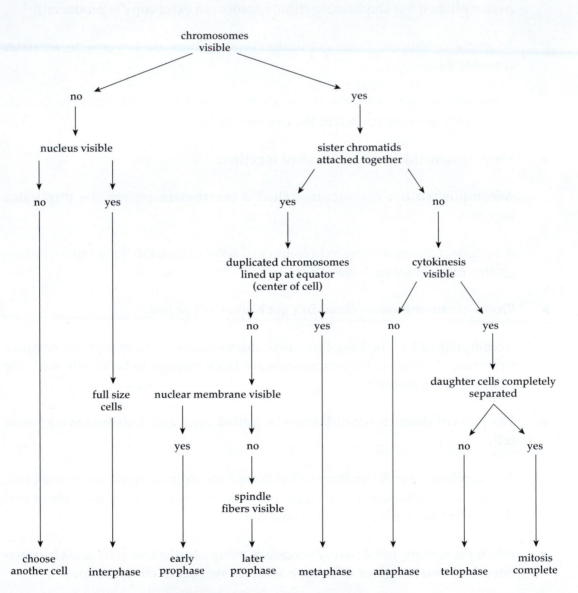

FIGURE 10-1.   Mitosis Decision Tree

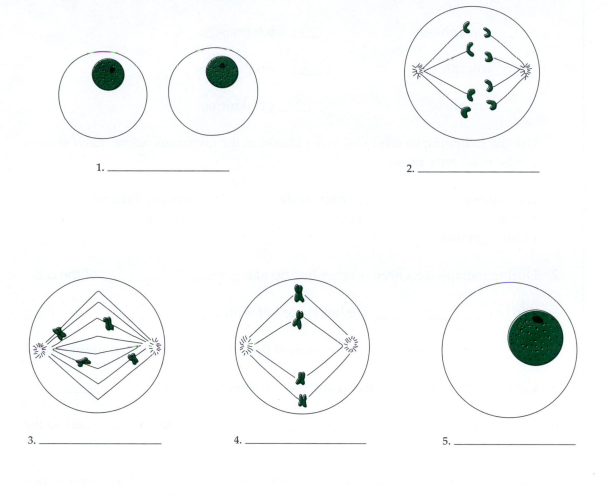

1. _____

2. _____

3. _____

4. _____

5. _____

6. _____

7. _____

FIGURE 10-2. Stages of the Cell Cycle to Be Identified

✓ Comprehension Check

1. Arrange the following stages of cell division in the correct order (1 through 6).

   _____ metaphase   _____ telophase

   _____ interphase   _____ prophase

   _____ anaphase   _____ cytokinesis

   Use the following words to fill in the blanks in the questions below. Each answer can be used **only once.**

   centromere   sister chromatids   cleavage furrow
   equator   cytokinesis   chromosomes
   mitotic spindle

2. During metaphase, chromosomes line up along the _____ of the cell.

3. The two _____ are held together with a _____.

4. _____ refers to the division of the cytoplasm into two daughter cells.

5. Genetic information in cells is contained in structures called _____.

6. During anaphase, part of the _____ can be seen attached to the centromere of each sister chromatid.

7. The constricted appearance of the cell membrane prior to the formation of two daughter cells is called the _____.

Check your answers with your instructor before you continue.

ACTIVITY 3   MITOSIS—THE REAL THING!

The stages of the cell cycle in living cells are often not as clear as in a set of diagrams. Sharpen your observation skills and practice recognizing the stages of mitosis at the same time.

1. For this activity, **work on your own.** Get the following supplies: **one slide of mitosis in whitefish embryos and a compound microscope.**

2. View the slide with the **high-power lens (40×).**

3. Your instructor will assign you **one stage of the cell cycle** to locate on your slide.

    When you locate a **clear** example of that stage, **indicate the cell with the pointer** in the ocular lens.

Check your identification with your instructor before you continue.

4. Draw the stage **as it appears under the microscope,** in the appropriate location on **Figure 10-3.**

    In your drawing, label the following (if present): **chromosomes, mitotic spindle, equator, cleavage furrow, cell membrane,** and **nuclear membrane.**

5. After completing your drawing, **observe** the other **stages of the cell cycle** as identified by other students.

    Create your own drawings of these stages and add them to the appropriate locations on **Figure 10-3.**

    In your drawings, label the following (if present): **chromosomes, mitotic spindle, equator, cleavage furrow, cell membrane,** and **nuclear membrane.**

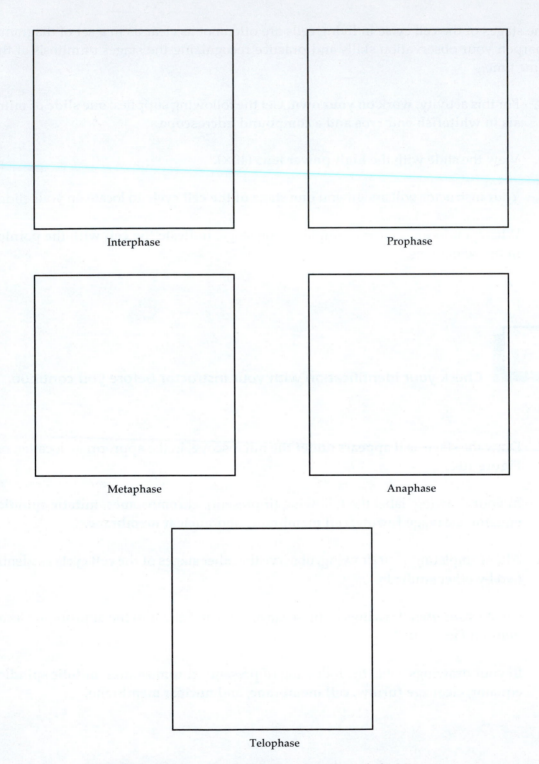

Interphase

Prophase

Metaphase

Anaphase

Telophase

FIGURE 10-3.   The Cell Cycle as Seen through the Microscope

ACTIVITY 4

When you look at a prepared slide, you're looking at a "moment frozen in time." The preserving chemicals stopped the cells in the middle of their daily activities. From this slide, you can see how many cells were undergoing cell division at that moment. You can even see how many cells were in each of the stages of mitosis. You can use this information to estimate the length of each stage of the cycle compared to the other stages.

1. For this activity, **work individually.** Get the following supplies: **one slide of onion root mitosis and a compound microscope.**

    View the slide with the **high-power lens.** Position your view of the onion root tip to a section where many cells are undergoing mitosis (see **Figure 10-4** for examples).

    Your instructor will give you additional instructions on how to position the slide for optimal viewing.

2. Using the stages of onion root tip mitosis in **Figure 10-4** as a guide, **count** the number of cells **in your field of view** that are in each stage of the cell cycle.

    Record your results in **Table 10-1.**

TABLE 10-1
RESULTS OF CELL CYCLE OBSERVATIONS

NUMBER OF CELLS IN STAGE	INTERPHASE	PROPHASE	METAPHASE	ANAPHASE	TELOPHASE
View 1					
View 2					
View 3					
Total for views 1–3					
Percentage of cell cycle	%	%	%	%	%

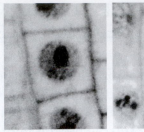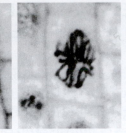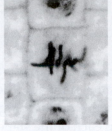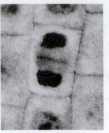

Interphase   Prophase   Metaphase   Anaphase   Telophase

FIGURE 10-4.   Stages of Mitosis in the Onion Root Tip

3.   Repeat the **instructions in Step #2 twice more** (so you'll examine a total of **three different onion fields of view**) and **record the results** for each in **Table 10-1**.

4.   **Total the number of cells counted** for each view and record the totals in **Table 10-1**.

5.   **Calculate** the percentage of cells observed in **each stage of the cell cycle.**

> STEP 1:
>
> Add the total number of cells counted in all five stages.
>
> Total cells counted = interphase + prophase + metaphase + anaphase + telophase
>
> Total cells counted = _____
>
> STEP 2:
>
> Divide the number of cells in interphase by the total from Step #1 and multiply by 100.
>
> $$\frac{\text{Number in interphase}}{\text{Total cells counted}} \times 100 = \underline{\hspace{2cm}} \%$$
>
> STEP 3:
>
> Repeat **Step #2** for the remaining four stages of the cell cycle. **Record all percentages** in **Table 10-1**.

6. Using the information in **Table 10-1,** answer the following questions about the **length** of each stage of the cell cycle.

   Which do you think is the longest stage? _____

   Which do you think is the shortest stage? _____

   Do you think any of the stages are about the same length? If so, list them below.

7. See **Table 10-2 (after the Self Test)** for the results of some experiments on the length of the onion cell cycle. Do the experimental results support your conclusions about the length of the various stages of mitosis? _____ **Explain** your answer.

ACTIVITY 5   REGENERATION

Some animals have the ability to replace (regrow) lost or damaged body parts. This process is known as **regeneration** and is a form of asexual reproduction. You may be familiar with some animals that have this ability. Starfish (also known as **sea stars**) can regenerate body parts as long as a small part of the central disk is present. Sea stars can grow a new arm and a severed arm can grow a whole new body! If cut in two, each half of the sea star can develop into a whole individual.

Planaria are freshwater flatworms that have the ability to regenerate lost body parts. Any piece of the body except the tail can regenerate a complete new adult. In this activity, you'll cut **planaria** to demonstrate regeneration and regrowth.

1. Work in groups. Get the following supplies: **one clean razor blade in a plastic holder, a glass petri dish, a pipette, and crushed ice.**

2. You'll also need a bottle of **pond water, labeling tape, and a dissecting microscope.**

    Set up your dissecting microscope, but keep the light **off.** On the stage of the microscope, place the **top half** of the petri dish.

    Fill the dish **to the rim** with crushed ice.

3. Place **one large planaria** into the **bottom half** of your petri dish with a small amount of **pond water.**

    Set the dish with the planaria **onto the crushed ice.** This will anesthetize the worm for the "operation."

    Let the worm relax on the ice for about **five minutes.**

4. While you're waiting, decide how you'll cut the planaria. You can choose one of the two cutting patterns in **Figure 10-5,** or develop your own pattern. However, **please note** that your cuts must be **perpendicular,** not at an angle.

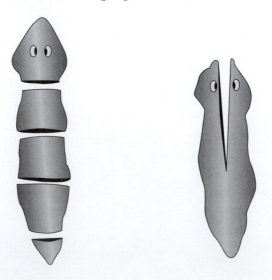

FIGURE 10-5. Planaria Cutting Patterns

5. With the razor blade, make **perpendicular cuts** through the worm. Don't tilt the razor blade when cutting. The cuts must be sharp and clean.

6. After the operation, add some clean **pond water** to the petri dish containing the worm.

    Discard the ice. Wash and dry the top half of the petri dish. Using labeling tape, **mark the lid** with **your laboratory section and your name.**

    Return the petri dish containing the worm to your instructor to be stored in a cool, quiet location.

7. After **two weeks,** you can examine your worm under the dissecting microscope to see the results of your experiment.

8. **Challenge Question!** If you cut a planaria into five pieces and each piece developed into a separate individual, would you be able to tell them apart? **Explain** your answer.

	ASEXUAL REPRODUCTION:
ACTIVITY 6	GROWING A PLANT FROM A CUTTING

As you've seen, mitosis provides genetically identical cells for growth, replacement, and repair of body tissues. Mitosis also provides a mechanism for **reproducing whole organisms asexually.** In asexual reproduction, offspring are produced that are **genetically identical** to the single parent and also to one another.

Plants don't have to grow from seeds. You can remove cuttings from your plants. The cuttings will develop roots and grow into new plants. This form of asexual reproduction is an example of **plant propagation.**

1. Get **a paper cup and a scalpel.**

    You'll also find **plants for propagation, rooting hormone, and potting soil.**

2. Fill the cup with **potting soil** up to **one inch** from the top of the container.

   Moisten the soil with water until it's damp, but not soggy.

3. **Cut a small piece of stem** with several leaves attached.

   Dip the **cut end** of the stem into the **rooting hormone** and **plant** it in the cup.

4. Keep your young plant **moist** and exposed to **bright, indirect light.** Roots should form within several weeks.

✔ Comprehension Check

1. A botanist is trying to save a rare plant from extinction. Suggest a method she can use to increase the population.

2. The **type of cell division** that produced root growth in my cutting is
   _____.

3. The possibility of human cloning, a type of asexual reproduction, is a frequently discussed topic. If a baby could be cloned using the DNA from one of her mother's skin cells, what percentage of the baby's traits would be similar to those of her mother? **Explain** your answer.

4. How would a colony of mice **produced by cloning be helpful** for testing the effectiveness of a new diet pill?

Check your answers with your instructor before you continue.

SELF TEST

Fill in the blanks with the most appropriate answer. Answers can be used **only once.**

a.   diploid
b.   sister chromatid
c.   centromere
d.   spindle fibers
e.   nuclear membrane
f.   cell membrane

g.   cytokinesis
h.   cleavage furrow
i.   equator
j.   zygote
k.   daughter cell
l.   chromosome

1. _____ Breaks down at the end of prophase, releasing the chromosomes.

2. _____ Chromosomes are arranged along this imaginary line during metaphase.

3. _____ Structure that holds two duplicate chromosomes together.

4. _____ Two of these are found at the end of mitosis.

5. _____ Normal number of chromosomes in a cell.

6. _____ Responsible for chromosome movement during mitosis.

7. _____ A constriction of the cell membrane that occurs in telophase.

8. _____ Duplicated chromosome has two of these.

9. _____ Process that separates the cytoplasm into two halves.

10. _____ Genetic material found in the cell nucleus.

Fill in the blanks with **true** or **false.**

11. _____ If a cell undergoing **mitosis** has a chromosome number of six, the daughter cells will have a chromosome number of three.

12. _____ In asexual reproduction, the offspring are identical to the parent.

13. _____ If cells are multiplying, as they do when grown in tissue cultures, mitosis is taking place.

14. _____ Growth during childhood is a good example of cell division.

15. Complete mitosis for the following parent cell. Draw each stage of mitosis **in order,** and show the number of chromosomes in each stage (including the two daughter cells). The beginning cell has a diploid number of four chromosomes.

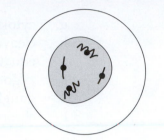

16. You're vacationing in Maine. One morning, while walking near the shore, you hear a group of fishermen complaining that there are too many sea stars in the ocean. They plan to solve this problem by netting the sea stars, cutting them in half to kill them, and throwing them back into the ocean. Is their plan likely to be effective? **Explain** your answer.

17. **Challenge Question!** If you wanted to asexually propagate a human, you could use

    a. a skin cell
    b. a sperm cell
    c. a circulating red blood cell
    d. any of the above would be suitable
    e. none of the above would work out

    Explain your answer.

TABLE 10-2

RESULTS OF EXPERIMENTS ON THE LENGTH OF STAGES IN THE ONION CELL CYCLE

	INTERPHASE	PROPHASE	METAPHASE	ANAPHASE	TELOPHASE
Length of stage	17 h	88 min	4 min	3 min	6 min
Percentage of cell cycle	91%	7.9%	0.4%	0.3%	0.5%

Connecting Meiosis and Genetics

Objectives

After completing this exercise, you should be able to:

- name and describe the stages of meiosis
- correctly use and understand the terminology associated with cell division and genetics
- demonstrate an understanding of the changes in chromosome number that occur during meiosis and fertilization
- compare meiosis I with meiosis II in terms of the position of the chromosomes in each stage, changes in chromosome number, and number of daughter cells produced
- explain the process and importance of crossing-over between homologous chromosomes
- draw and complete Punnett squares and use them to determine genetic probabilities in monohybrid crosses

CONTENT FOCUS

Your body cells contain **46 chromosomes** within the nucleus. As you probably know, **half** of this genetic information (23 chromosomes) is inherited from your mother and **half** (the other 23 chromosomes) from your father.

The genetic information carried by the **gametes** (the sperm and egg), when incorporated into a fertilized egg, will determine all the physical and physiological traits of the offspring.

There is a special type of cell division that changes the chromosome number from the normal **diploid** number (46) to the **haploid** number (23) found in sperm and eggs. This special type of cell division is called **meiosis.**

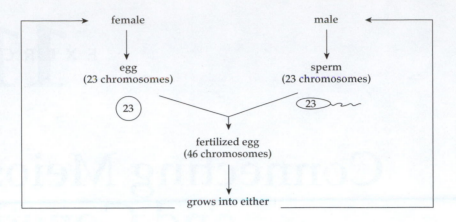

The stages of meiosis, in most respects, are similar to those of mitosis. The important differences include the following:

- the chromosome number is reduced from diploid to haploid to form gametes
- an exchange of genetic material takes place in a process known as **crossing-over;** gametes produced by meiosis are **not** genetically alike
- meiosis involves two cell divisions (meiosis I and meiosis II); one occurs immediately after the other

The genes on a chromosome may exist in **more than one form,** called **alleles.** Individuals inherit **two alleles for each trait,** one received from the mother in the egg and the other from the father in the sperm.

If the two inherited alleles represent **different** forms of the gene (**different** alleles), an individual is called **heterozygous for that trait (Bb).** The allele that is **expressed in heterozygous individuals** is referred to as **dominant** and is represented by a capital letter **(B).**

The allele **not expressed** in heterozygous individuals is called **recessive.** Recessive alleles are represented with lowercase letters **(b).**

If individuals inherit two identical alleles for a trait **(BB or bb),** they are said to be **homozygous for that trait.**

The **combination of alleles** for a trait represents an individual's **genotype** (such as BB or Bb).

The **physical description** of a specific trait is called the **phenotype** (such as brown eyes, blood type, or freckles).

In this exercise, you'll work through the process of meiosis to **form** male and female **gametes, decode** the genetic information carried by these gametes, and use that information to **build** your baby's face.

GETTING STARTED:
ACTIVITY 1   BUILD A PAIR OF CHROMOSOMES

1. Work in groups. Get the following supplies: **a bag labeled "diploid human genome, male" or "diploid human genome, female."**

   In the supply area, you'll find containers of **beads** of several colors and shapes (representing **alleles**) and magnetic **connectors** (representing **centromeres**).

   Although humans have 23 pairs of chromosomes, we'll simplify the process by using only **one chromosome pair.**

2. Fill your bag with **four beads from each jar.** Note that there are two jars of beads of each color. In one jar, the beads are striped. In the other jar, they're solid.

   Solid-colored beads represent **dominant alleles** and **striped beads** represent **recessive alleles for the same trait.**

   Get **two red centromeres and two yellow centromeres.**

3. **Remove Table 11-1 from one group member's book and lay it flat on the laboratory table. As you draw each bead (representing the alleles of your maternal and paternal chromosomes) from the bag, place it into the appropriate box in Table 11-1.**

   Bead selection is random! Don't look in the bag while you're drawing the beads.

   a. The first bead you draw will become part of the chromosome you inherited from your **mother** (**maternal** chromosome).

   b. The next bead of the **same color** you draw will become part of the chromosome you inherited from your **father** (**paternal** chromosome).

   c. Additional beads of the **same color** you draw will be **THROWN BACK INTO THE BAG.**

   d. Draw beads until you've drawn **one of each color or shape for the maternal chromosome and one of each color or shape for the paternal chromosome.**

4. When you've drawn all the beads you need for both chromosomes, you're ready to hook your alleles together.

   Take a **yellow centromere** and hook it into position **between the green and red beads** of your **maternal chromosome.**

   Keeping the beads in the **correct order,** attach the remaining beads to the chromosome.

TABLE 11-1
TRAITS ON YOUR CHROMOSOMES

Bead Color or Shape	Maternal Alleles	Paternal Alleles	Trait	Alleles (Genotype)
Purple			Face shape	FF or Ff—round ff—triangular
Orange			Hair texture	HH—curly Hh—wavy hh—straight
Yellow			Eye size	EE—large Ee—medium ee—small
Blue			Eye distance	DD—close together Dd—medium spacing dd—far apart
Green			Eyebrow shape	BB or Bb—thick bb—thin
Red			Eyelash length	LL or Ll—long ll—short
White			Nose size	NN—big Nn—medium nn—small
Pink			Lips	GG or Gg—thick gg—thin
Black			Ear lobes	RR or Rr—free rr—attached
White oval			Cleft in chin	TT or Tt—present tt—absent
White twisted			Freckles	QQ or Qq—freckles qq—no freckles

5. **Repeat step #4** to connect the alleles of the **paternal chromosome** using the **red centromere.**

   Together these **two chromosomes (maternal and paternal)** are referred to as **homologous chromosomes.** Homologous chromosomes carry alleles for the **same traits** (face shape, eye size, etc.), although the genetic information is **not identical.**

6. Fill in **Table 11-2 with the genotypes and phenotypes of the traits on the chromosomes** you've just constructed.

   Solid beads = dominant alleles. Striped beads = recessive alleles.

T A B L E 1 1 - 2
WHAT ARE YOUR TRAITS?

TRAIT	YOUR GENOTYPE	YOUR PHENOTYPE
Face shape		
Hair texture		
Eye size		
Eye distance		
Eyebrow shape		
Eyelash length		
Nose size		
Lips		
Ear lobes		
Cleft in chin		
Freckles		

Note:

When selecting your genotypes and phenotypes from Table 11-1, you probably noticed something unusual. Some of the traits have two possible phenotypes (round face or triangular face), and others have three options (curly hair, wavy hair, or straight hair).

The traits with three possible phenotypes represent a different mode of inheritance called **incomplete dominance.** Incomplete dominance is a special type of inheritance that occurs when an allele exerts only partial dominance over another allele. This results in a third, intermediate phenotype in heterozygous individuals.

HOMOLOGOUS CHROMOSOMES
ACTIVITY 2   SEPARATE IN MEIOSIS TO FORM GAMETES

1. Draw a series of circles that are the same as those shown in **Figure 11-2** on page 182.

   Make the circles big enough for your chromosomes to fit comfortably.

2. Place your pair of chromosomes in the **first** circle. Your cell is now in **interphase.**

   The diploid number of chromosomes in our simulated cell is 2.

   How many chromosomes came from your mother? _____

   How many from your father? _____

   How many **traits** are represented on your chromosomes? _____

   How many **alleles** are represented on your chromosomes? _____

3. An important event that occurs during **interphase** involves the replication of chromosomes. Use the spare beads in your genome bag to **replicate** your two chromosomes.

   Two identical (replicated) DNA strands are called **sister chromatids.**

   Attach the new **sister chromatids** for your **maternal chromosome** together at the centromeres. Do the same for your replicated **paternal chromosome.**

4. Move your duplicated chromosomes to the next circle, marked **prophase** through **telophase of meiosis I.**

   You'll complete **prophase through telophase of meiosis I** in the **same circle.**

5. As shown in **Figure 11-1,** during **prophase I,** the **two homologous chromosomes** find each other and pair up in a process called **synapsis.**

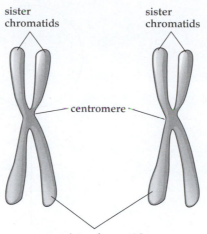

FIGURE 11-1. Pattern for Meiosis Simulation

6. During synapsis, the **non-sister chromatids** exchange genetic information. This process is called **crossing-over.**

   Simulate crossing-over between **two non-sister chromatids** in your homologous pair (exchange of alleles between **one maternal and one paternal** chromosome).

   Exchange alleles for the **last five traits** (nose size through freckles).

Note:

Exchange alleles ONLY for two NON-SISTER chromatids!

---

7. Simulate **metaphase I** by placing the chromosomes **in the correct position** relative to the **equator.**

   To simulate **anaphase I,** separate the two homologous chromosomes by moving them to the opposite poles of the cell.

   To simulate the division of the cytoplasm in **telophase I,** draw a dotted line that represents the separated **daughter cells.**

interphase

prophase through telophase
of meiosis I

meiosis I daughter cells
and prophase through
telophase of meiosis II

meiosis II daughter cells

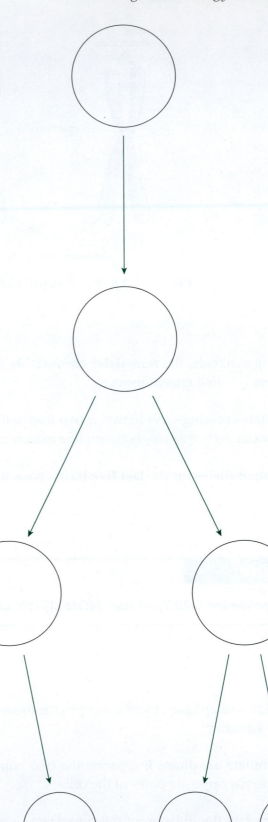

FIGURE 11-2.   Duplicated Homologous Pair (Tetrad)

8. Move your chromosomes to the next two circles that are labeled **meiosis I daughter cells.**

✓ Comprehension Check

1. What is the number of chromosomes in each daughter cell? _____

2. Does this number represent the **diploid** or the **haploid** chromosome number? _____

3. What happened to the **maternal and paternal** chromosomes from the original parent cell?

    Are the two daughter cells genetically identical? _____ **Explain** your answer.

---

9. Meiosis continues with a second cellular division.

    Simulate meiosis II **without moving your chromosomes** to another set of circles. Note that the circles you're now using have two labels. In **addition** to meiosis I daughter cells, the circles are also labeled **prophase through telophase of meiosis II.**

    Nothing unusual occurs to the chromosomes in prophase II.

    To simulate **metaphase II,** place the chromosomes **in the correct position** relative to the **equator.**

    To simulate **anaphase II,** separate the two **sister chromatids** by moving them to the opposite poles of the cell.

    Move your chromosomes to the four circles that are labeled **meiosis II daughter cells.** Each of these daughter cells has the potential to develop into a sperm or egg.

Check the movement of your chromosomes through meiosis I and II with your instructor before you continue.

✓ Comprehension Check

1. The original parent cell in your meiosis simulation had two alleles for each trait.

   How many alleles for each trait are **now** in each daughter cell? _____

2. Are your daughter cells diploid or haploid? _____

3. Are the four daughter cells **genetically identical?** _____ **Explain** your answer.

4. Where in the body does meiosis occur in **males?** _____

   Where does it occur in **females?** _____

Check your answers with your instructor before you continue.

ACTIVITY 3   FERTILIZATION: NATURE'S EQUIVALENT TO ROLLING THE DICE

1. As you know, **only one sperm and one egg** can participate in fertilization.

   Any of the four sperm can potentially fertilize an egg, but the development of female gametes is slightly different. **Only one** of the four daughter cells will develop into an egg cell. The other three aren't functional and are referred to as **polar bodies.**

   You'll simulate fertilization of your gametes with a roll of the dice. Get a **single six-sided die.**

   Are **your** gametes potential sperm or eggs? _____

2. Choose the gamete that will participate in fertilization as follows:

   a. **Number your daughter cells** 1 through 4.

   b. **Roll the die.** If you get a number between 1 and 4, that's your lucky gamete. If you get 5 or 6, roll again.

3. Link up with a group that has produced a gamete of the opposite sex. Take the chromosomes from both gametes and **place them in a circle** labeled **"Fertilized Egg."**

ACTIVITY 4 WHAT'S YOUR BABY'S GENOTYPE?

1. On the two chromosomes in **Figure 11-3,** list your baby's **alleles in their correct order.**

   Solid beads = dominant alleles. Striped beads = recessive alleles.

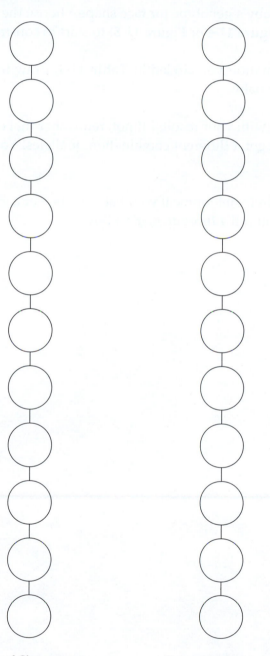

Maternal Chromosome   Paternal Chromosome

FIGURE 11-3. Maternal and Paternal Chromosomes of the Fertilized Egg

ACTIVITY 5   GENOTYPE DETERMINES PHENOTYPE

1. Now that you know your baby's **genotype** for each trait, turn back to **Table 11-1.** In **Table 11-1, highlight or circle** your baby's **genotype and phenotype** for each trait.

    Now that you know your baby's **phenotype** for each trait, you're ready to see what your baby will look like!

2. Based on your baby's genotype for face shape, choose the appropriate face shape outline (either **Figure 11-4** or **Figure 11-5)** to start to construct your baby's face.

    Using the information you circled in **Table 11-1,** draw in the appropriate facial features for your baby.

    Are you pleased with your results? If not, remember that each time a sperm fertilizes an egg, you get a different combination of alleles. You may have better luck next time.

3. **Roll the die again** to determine if your baby is a boy or a girl. If you roll **1 through 3,** it's a **girl.** If you roll **4 through 6,** it's a **boy.**

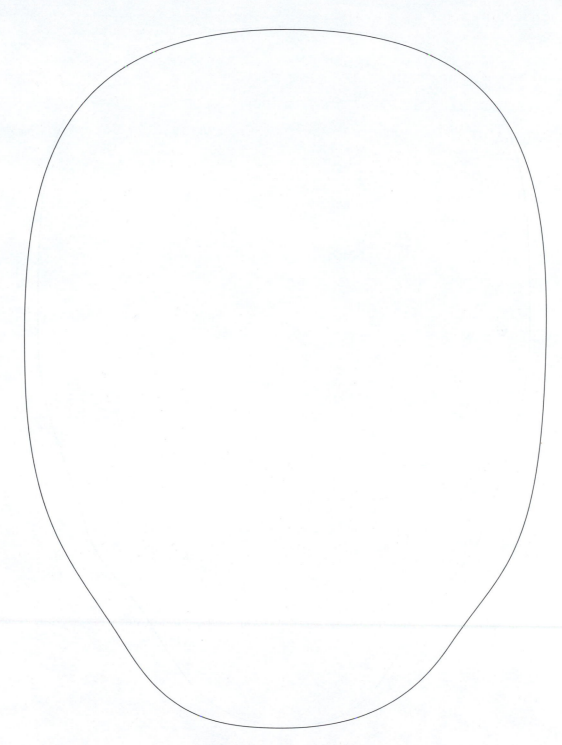

FIGURE 11-4. Round Face

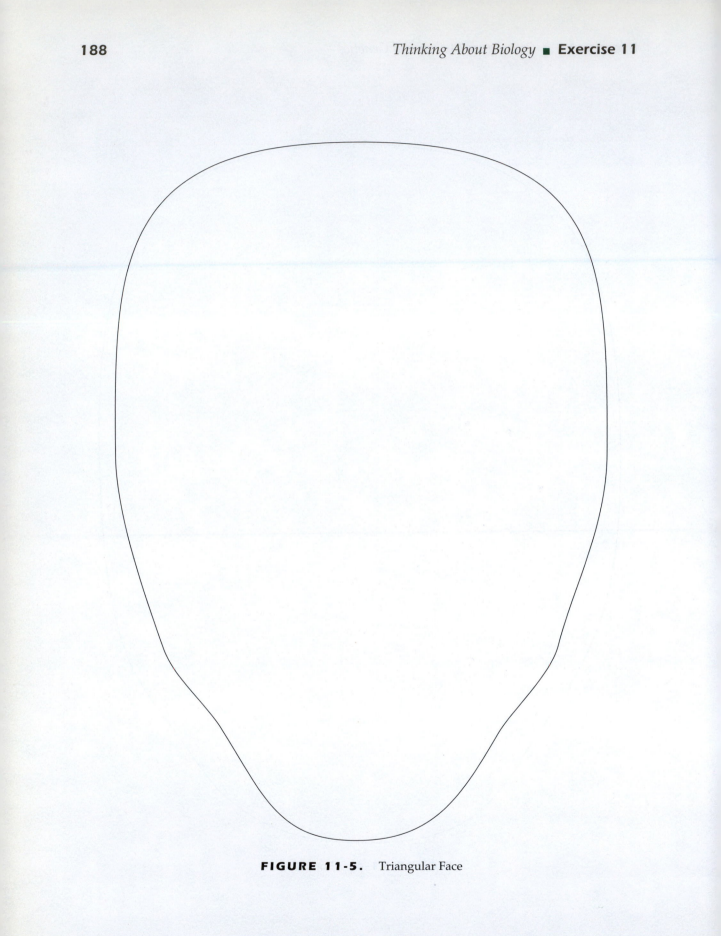

FIGURE 11-5.   Triangular Face

ACTIVITY 6   PASSING ON TRAITS

Twenty-four years have passed. Your baby is now an adult and has married someone who's **heterozygous for freckles.** What is the probability that they'll have a child with freckles?

The first step in calculating the probability of freckles is to **determine the alleles carried by the sperm and eggs** of this couple.

As you know, alleles separate in meiosis, so each sperm and each egg has only **one allele from each homologous pair.**

Your child's genotype for the freckle trait _____ _____

1. A **Punnett square** is a convenient way to visualize the different combinations of alleles that might occur during fertilization.

2. The spouse's alleles have been entered on the top row of the Punnett square. Enter the alleles for **your child** along the left side of the square.

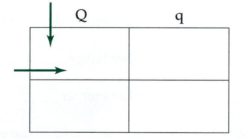

3. Using the Punnett square as a guide, fill in the boxes by carrying the alleles down from the top and across (from left to right).

    Each box in the Punnett square represents a **25% probability** of having a baby with that specific genotype.

    What's the probability that this couple will have a child with freckles? _____

ACTIVITY 7   PRACTICING GENETICS PROBLEMS

1. Nicole Johnson is an **albino,** a condition inherited by a **recessive allele (a).** She's marrying a man with normal skin color, whose father was an albino. Nicole makes an appointment for genetic counseling because she wants to evaluate the risk of passing her condition on to her children. What is the probability that a child from this marriage will be an albino? **Show your work.**

2. In humans, assume that **brown eyes (B) are dominant over blue eyes (b).** A blue-eyed man whose father was brown eyed and whose mother was blue eyed marries a brown-eyed woman whose father had brown eyes and whose mother had blue eyes. What are the genotypes of all the individuals? **Show your work.**

    blue-eyed man _____   brown-eyed woman _____

    his father _____   her father _____

    his mother _____   her mother _____

    What is the probability that this couple will have a blue-eyed child? _____

    What is the probability of a brown-eyed child? _____

3. **Polydactyly** (extra fingers or toes) is inherited by a **dominant allele (F).** A father is polydactyl, the mother has the normal number of fingers, and they have had one normal child. What are the genotypes of all the individuals involved? What is the probability that a second child will have a normal number of fingers? **Show your work.**

   the father _____   the mother _____

   the first child _____

   probability of a child with five fingers _____

4. Assume that the baby you made in **Activity 5** has **wavy hair,** inherited through **incomplete dominance (Hh)** and grows up to marry a person who also has **wavy hair.**

   What is the probability you'll have grandchildren with **straight** hair? _____

   Curly hair? _____

   Wavy hair? _____

   Show your work.

SELF TEST

1. Mark each statement below as **true (T) or false (F).**

_____ Sperm cells have the same number of chromosomes as the man that produced them.

_____ A zygote is a fertilized egg.

_____ Crossing-over occurs in both mitosis and meiosis.

_____ Gametes are genetically identical to each other.

_____ The letters AA are used to indicate a heterozygous phenotype.

_____ Sister chromatids are identical before crossing-over occurs.

_____ Sister chromatids are held together by a structure called synapsis.

_____ Daughter cells produced by meiosis are haploid.

_____ When the chromosomes line up during metaphase I of meiosis, the
equator separates the maternal and paternal members of each pair.

_____ The daughter cells produced by meiosis have both a maternal and a
paternal chromosome from each homologous pair.

2. Can a sperm cell contain maternal chromosomes? **Explain** your answer.

3. Why are chromosomes sometimes drawn as a straight line and other times in an
"X" configuration?

4. **Albinism** is a **recessive** condition **(aa)** in which body cells can't manufacture the pigment **melanin,** which colors eyes, hair, and skin. **Normal pigment production is dominant (AA or Aa).** A normally pigmented woman marries a normally pigmented man. To their surprise, they have an albino child. Give the genotypes of the parents and the child:

   the father _____   the mother _____   the child _____

   What is the probability that this couple will have another albino child? _____

   What is the probability that they'll have a normally pigmented child? _____

   Show your work.

5. **Tay-Sachs disease** in humans is controlled by a **recessive allele (t).** This disease is characterized by the inability to produce an enzyme needed to metabolize lipids in brain cells. Without this enzyme, lipids accumulate in the brain cells and gradually destroy the ability of the cells to function. Children affected with this disease usually die by age five.

   What genotypes **must** be found in **both parents** in order to have a child with Tay-Sachs? _____   _____

   Explain your answer and **include a Punnett square** that shows the possible genotypes among the children of this marriage.

6. Imagine you're a zookeeper in the Big Cat house of the National Zoo in Washington, D.C. Among the tigers, two yellow-coated parents (Ghandi and Sabrina) give birth to a cub named Snowflake that has a rare color variation—a white coat with black stripes.

    Assuming that the **white coat is inherited as a recessive allele (c),** what are the genotypes of the three tigers?

    Ghandi ＿＿＿＿＿＿＿   Sabrina ＿＿＿＿＿＿＿   Snowflake ＿＿＿＿＿＿＿

    If these parents have another cub, what is the probability that the cub will be **heterozygous** for the coat-color trait? **Show your work.**

7. Another zoo wants to start its own tiger exhibit. So far, the zoo has one tiger in its collection, a white one. If the National Zoo sends Ghandi on breeding loan to this zoo, what is the probability that the zoo will get another white tiger for its new exhibit? **Show your work.**

Human Genetics

Objectives

After completing this exercise, you should be able to:

- demonstrate an understanding of the limitations of sample size in scientific data analysis
- determine genotypes using pedigree charts
- explain why more males than females express X-linked traits
- solve genetics problems involving dominant-recessive inheritance, X-linked traits, multiple allele traits, and codominance
- explain the evolutionary relationship between sickle-cell disease and malaria
- apply your knowledge of genetics to real-life situations

CONTENT FOCUS

Not all alleles produce visible traits such as skin color or height. Most alleles control **physiological** traits, such as production of digestive enzymes, hormones, and antibodies. Alleles are responsible for invisible traits such as blood type, ability to carry out metabolic pathways (such as producing proteins or storing blood sugar), color vision, and many others. One example of a physiological trait controlled by a **single gene** is the ability to taste a harmless chemical, **PTC** (phenylthiocarbamide).

The story of the discovery of the "bitter taste gene," which controls the ability to taste PTC, is quite interesting. It happened in 1931 at Dupont Chemical Company. Some PTC crystals accidentally blew into the air in the room, and into the mouths of scientists working in that lab. Some of the scientists complained of a bitter taste, but others said the chemical had no taste! After this observation, the scientific method took over. The scientists tested friends, family members, and coworkers, and found similar results. Some tasted PTC as bitter and others couldn't taste it at all. Seventy years later, these results were linked to a single gene located on chromosome #7.

195

ACTIVITY 1 PTC TASTING

1. Get the following supplies: **one piece of control taste paper and one containing PTC.**

2. **Taste the control paper** first (to establish the taste of the paper itself). Then **taste the PTC paper.**

    If you're a taster, you'll detect a very unpleasant, bitter taste. If there's little difference between the PTC and the control papers, you're not a taster.

    Are you a taster or a nontaster? _____

3. Record your taste-test results **on the master chart in the room.**

4. In the general population, approximately 75% of people can taste PTC. The remaining 25% aren't able to taste this chemical.

    Were your class results close to the **expected percentage?** _____

    If not, suggest a possible reason why your class results differed from the expected outcome:

5. The ability to taste PTC comes from a **dominant allele (T).** Using this information, fill in the appropriate genotypes for tasters and nontasters.

    Tasters _____   _____

    Nontasters _____

6. The inheritance pattern of traits such as PTC tasting can be diagrammed in a chart called a **pedigree.** A pedigree illustrates the marriages for several generations within a family and the children produced.

    ■ **Females** are shown with **circles** and **males** are shown with **squares.**

    ■ A **black square or circle** shows the **presence of the condition** being studied. A **white square or circle** means the **condition is absent** in that person.

    ■ A marriage or mating is shown by a line connecting the parents.

    ■ Children from a mating are shown by a vertical line between the parents.

    ■ All individuals from the **same generation** are shown along the same horizontal line.

7. After examining the **pedigree key, determine the genotypes of all the people in this family.** In a few cases, there may not be enough information to determine a person's second allele. In this situation, **enter a question mark (?) in place of the second letter.**

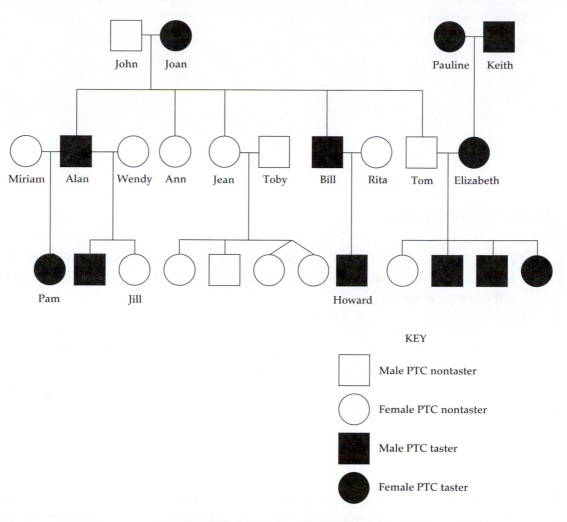

KEY

☐ Male PTC nontaster

◯ Female PTC nontaster

■ Male PTC taster

● Female PTC taster

FIGURE 12-1. PTC Pedigree

8. Referring to the pedigree in **Figure 12-1,** if Jill marries Howard, what is the **probability** that their children will be

tasters? _____   nontasters? _____

Show your work!

9. If Pam marries Howard, what is the probability that their children will be

   tasters? _____   nontasters? _____

   Show your work!

 Comprehension Check

1. After performing the PTC experiment, do you think that the ability to taste cer-
    tain chemicals can explain why your children's food preferences might be differ-
    ent from your own?

2. **Barium sulfate** is tasteless to some people, but bitter to others. Two sisters
    are having their gastrointestinal tracts x-rayed. They are both asked to drink a
    barium sulfate "milk shake" before the procedure. One sister drinks the milk
    shake without complaint. The other sister complains that it tastes terrible. Using
    the information gained in today's lab activities, **suggest an explanation for this
    difference.**

Check your answers with your instructor before you continue.

ACTIVITY 2   X-LINKED TRAITS

In humans, **the X chromosome is large in comparison to the Y chromosome.** The X chromosome carries information for many traits that aren't related to the sex of the individual. Alleles carried only by the X chromosome are said to be **X-linked** (or sometimes, sex-linked).

Some of the alleles on the tiny **Y chromosome** appear to have no counterparts on X. These **Y-linked alleles** code for traits that are found **only in males.**

Among the X-linked traits are a number of recessive genetic disorders. One of these is **hemophilia,** the inability to produce proteins necessary for blood clotting. Hemophiliacs can bleed to death from relatively minor cuts or bruises. Historical records dating back thousands of years mention the inheritance pattern of hemophilia. Among the ancient Hebrews, sons born to women with a family history of hemophilia were excused from circumcision.

Hemophilia was common during the 1800s in the royal families of Europe, whose members often intermarried. Queen Victoria of England was a **carrier** of the trait. She had one X chromosome with the allele for normal blood clotting (X^H) and the other with the defective allele (X^h). Because she did have **one normal dominant allele,** her blood clotted normally.

Her husband, Prince Albert, was completely normal for this trait. He had one normal allele on the X chromosome (X^H) and a Y chromosome with **no allele related to blood clotting** (Y^o).

Eighteen of Queen Victoria's 69 descendants were carrier females or hemophiliac males. Crown Prince Alexis of Russia was one of these hemophiliac descendants. His affliction indirectly contributed to the overthrow of the monarchy in Russia.

1.   Complete this Punnett square for the marriage of Victoria and Albert.

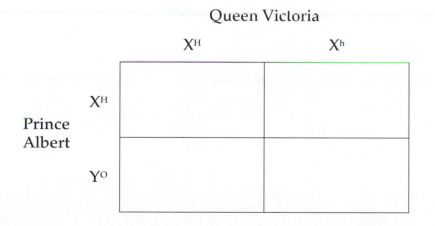

2. Considering the **entire Punnett square,** what is the probability that Victoria and Albert could have

   a hemophiliac son _____

   a hemophiliac daughter _____

   a normal son _____

   a carrier daughter _____

   a daughter with normal blood clotting _____

3. A partial pedigree of hemophilia in the descendants of Victoria and Albert is shown in **Figure 12-2. Affected individuals are indicated with darkened circles or squares.** For **each individual in the pedigree,** fill in his/her probable genotype.

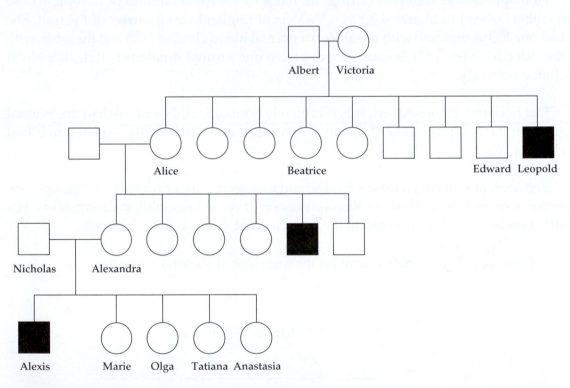

FIGURE 12-2.   Partial Pedigree of Hemophilia in the Descendants of Victoria and Albert

✔ Comprehension Check

1. **Challenge Question!** If Victoria and Albert had a hemophiliac daughter, why would it be **unlikely for her to survive puberty** without medical intervention?

ACTIVITY 3   CODOMINANCE AND SICKLE-CELL DISEASE

Now that you're becoming an expert in solving genetics problems, let's try something a little more challenging. **Sickle-cell disease** is an inherited trait controlled by a pair of alleles. The allele for normal hemoglobin is designated as Hb^A. Inheritance of one or more mutated Hb^S alleles for the hemoglobin gene alters the folded protein structure of the hemoglobin molecules produced in red blood cells. Red blood cells filled with mutated hemoglobin molecules take on an abnormal crescent shape (see **Figure 12-3**). The term "sickled" comes from an agricultural tool that has a curved, sharpened blade to harvest grain.

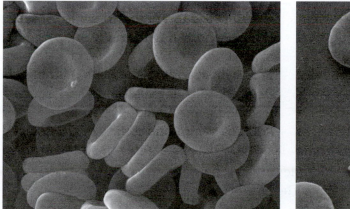

a. Normal Red Blood Cells

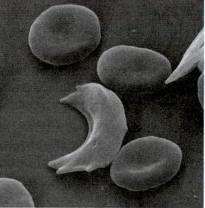

b. Sickled Red Blood Cells

FIGURE 12-3.   Comparison of Normal and Sickled Red Blood Cells

Sickle-cell disease is inherited through codominance. **Codominance** is a special type of inheritance in which **two alleles are equally dominant.** Both alleles are **expressed independently of each other** (referred to as **sickle-cell trait**). A person with sickle-cell trait is a heterozygous individual who shows **both** homozygous phenotypes and has **both** normal and sickled red blood cells in circulation.

Sickle-cell anemia, a blood disorder affecting transport of oxygen by hemoglobin, is a serious medical condition. The sickled cells carry less oxygen than normal red blood cells and also block capillaries, depriving the tissues of the oxygen they need. Individuals with sickle-cell anemia frequently die at a relatively early age.

Individuals with sickle-cell trait are usually healthy but experience some problems with intense exercise or under low-oxygen conditions. In the United States, sickle-cell trait affects 1 out of 12 African Americans.

The following is a summary of the possible sickle-cell genotypes and phenotypes:

Hb^AHb^A completely normal

Hb^AHb^S sickle-cell trait (this person has a combination of normal hemoglobin and the abnormal, sickled form of hemoglobin)

Hb^SHb^S sickle-cell anemia (all abnormal hemoglobin)

Sickle-cell disease is quite common in countries that have a high incidence of malaria. Work the following problems to determine the probabilities for inheritance of the sickle cell allele, and in Activity 4, you'll determine why a relationship exists between sickle cell disease and malaria.

1. If **both parents are heterozygous** for sickle-cell disease, what are the possible genotypes and phenotypes for their children?

2. What is the probability of a couple having a child with sickle-cell **trait** if **one** parent is **normal** and the **other** has **sickle-cell trait?**

ACTIVITY 4   A FAMILY HISTORY OF SICKLE-CELL DISEASE

The foundation of our modern knowledge of sickle-cell disease is based on the research of Dr. Angela Ferguson and Dr. Roland Scott, professors at Howard University in Washington, D.C. The doctors became interested in sickle-cell disease because many of their friends and some family members were affected by it. They published the first research paper on sickle-cell disease in the 1940s—**25 years ahead** of other researchers.

Imagine that you're a physician studying blood genetics in the laboratory of Drs. Ferguson and Scott. The Minister of Health from Nigeria has requested their help to investigate two health problems within the villages of his country that seem to be related: sickle-cell disease and malaria. Drs. Ferguson and Scott send you to investigate.

After spending a year looking into the problem, you've collected data about **Family A** and constructed a **pedigree** (see **Figure 12-4** on page 204) **and a health profile** for this family (see **Table 12-1**). **Each individual in the pedigree is identified by a number.**

TABLE 12-1
HEALTH PROFILE OF FAMILY A

HEALTH STATUS	AFFECTED FAMILY MEMBERS
Currently have malaria	13, 32, 33
Dead from malaria	7, 9, 10, 17, 29, 30, 31, 34
Dead from sickle-cell disease	8, 15, 25, 26, 35
Dead from causes unrelated to malaria or sickle-cell disease	3, 19
In good health	1, 2, 4, 5, 6, 11, 12, 14, 16, 18, 20, 21, 22, 23, 24, 27, 28, 36, 37, 38

Is there any relationship between sickle-cell disease and malaria in Family A? If there is a relationship, why would this occur? To answer these questions, you need some background information about malaria.

Malaria is an infection of the blood carried from person to person by mosquitoes. People of all ages can be infected with malaria, but babies, young children, and pregnant women suffer the worst effects. In 2009, the World Health Organization recorded 225 million malaria cases and 781,000 deaths. Although efforts to prevent infection through mosquito netting, pesticide spraying, and vaccine have dramatically increased, the problem is still severe. Most deaths occur in children in sub-Saharan Africa.

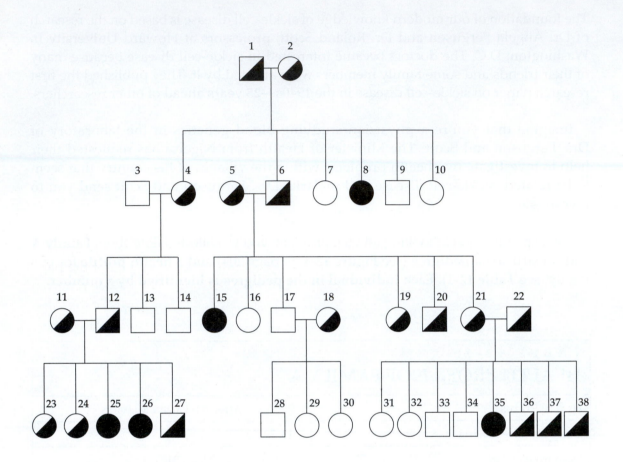

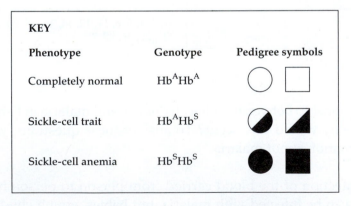

FIGURE 12-4. Pedigree of Family A

Malaria is caused by protozoans (one-celled organisms) of the genus *Plasmodium*.

- Infection begins with a bite from an infected mosquito.

- The parasite travels from the mosquito to the person's liver, where it begins to reproduce.

- The parasite leaves the liver and travels to the bloodstream, where it infects the **red blood cells.** The parasite reproduces in the red blood cells, which destroys the cells and releases more parasites into the bloodstream.

- The disease causes red blood cells to burst every 48 hours, releasing toxins that cause fever, chills, and even death.

- If another mosquito bites an infected person, that mosquito can then carry the infection to someone else.

Because malaria parasites reproduce inside red blood cells, the altered shape and oxygen-carrying capability of sickled cells decrease the reproductive rate of the parasites. In addition, when infected blood cells pass through the spleen, they're removed and destroyed (along with the malaria parasites). This reduces the severity of the infection. Therefore, people with sickle-cell trait have a selective advantage for survival in countries where malaria is prevalent.

1. Refer to the Health Profile of Family A in **Table 12-1.**

    On the pedigree in **Figure 12-4,** place the letter "**M**" next to the numbers of all persons who **currently have malaria** or who have **died from malaria.**

    Place the letter "**H**" next to the numbers of all family members who are currently **in good health.**

    What is the most common genotype for people in Family A who currently have or who have died from malaria? _____

    Among the **healthy individuals** of Family A, what is the most frequent genotype? _____

2. In **Table 12-2,** enter the number of deaths in **Family A** from each cause (as shown in Table 12-1). Calculate the percentage who died from each cause and record your results in Table 12-2.

    What is the leading cause of death among the members of Family A? _____

TABLE 12-2 CAUSES OF DEATH IN FAMILY A COMPARED TO NEIGHBORING COMMUNITIES				
CAUSES OF DEATH	FAMILY A NUMBER DEAD	FAMILY A % OF TOTAL DEATHS	NEIGHBORING COMMUNITIES NUMBER DEAD	NEIGHBORING COMMUNITIES % OF TOTAL DEATHS
Malaria			1080	
Sickle-cell disease			600	
Other			320	
TOTAL			2000	

3.   Return to **Table 12-2** and calculate the percentage of total deaths in the neighboring communities.

     In **Figure 12-5,** plot a graph comparing the **percentage of deaths from each cause** in **Family A** compared to the **neighboring communities.** Place the appropriate labels on the X and Y axes and enter a title for the graph.

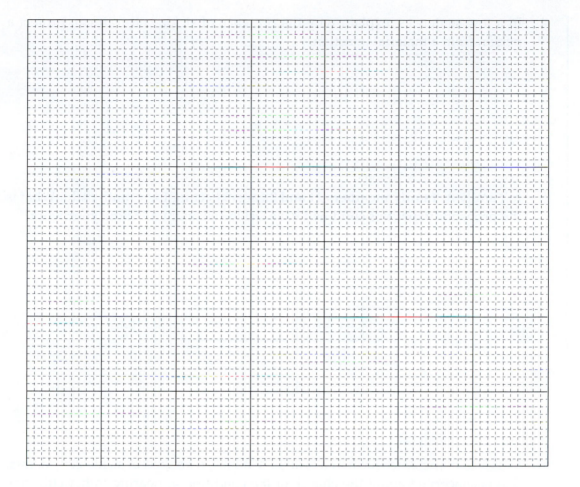

FIGURE 12-5. _____

Check your graph with your instructor before you continue.

Comprehension Check

1. Is the death pattern shown by Family Λ typical of the population as a whole?

2. What causes malaria?

3. What causes sickle-cell disease?

4. In a few sentences, explain why the presence of sickle-cell trait can protect a person from a malarial infection.

5. In Family A, **family members 21 and 22** are expecting their ninth child. What is the probability that this child will have sickle-cell anemia? **Show your work!**

<table>
<tr><td></td><td></td></tr>
<tr><td></td><td></td></tr>
</table>

6. Closer to home, Michelle had a brother, Charles, who died of sickle-cell anemia. She is concerned about the chance of the condition appearing in her children. When blood samples were taken and placed under low-oxygen conditions, some of her red blood cells sickled. Those of her husband James, however, remained normal when tested. **Show your work and list the genotypes of all those mentioned in the problem.**

    Michelle _____   Charles _____   James _____

    What is the probability that Michelle's and James's children will have sickle-cell anemia? _____

    Sickle-cell trait? _____

7. **Challenge Question!** Tanya, who has the genotype Hb^AHb^S, and her sister Amelia, who is Hb^AHb^A are planning a vacation to visit the ruins of a famous Inca city high in the mountains of Peru. Should either Tanya or Amelia be concerned about taking this trip in view of their sickle-cell status? **Explain** your answer.

Check your answers with your instructor before you continue.

ACTIVITY 5   CODOMINANCE AND MULTIPLE ALLELES

Up to this point, we've dealt with traits that have **only two alleles.** Human blood types (A, B, and O) are inherited by **multiple alleles.** There are **three possible alleles** for blood type. **Two of the three** possible alleles are **codominant.**

Codominance is a special type of inheritance in which two alleles are equally dominant. Both alleles are expressed independently of each other, resulting in a heterozygous individual who shows both homozygous phenotypes.

The four major blood groups are determined by the presence or absence of two antigens, referred to as antigen A and antigen B. The antigens are part of the cell membrane of red blood cells (see **Table 12-3**).

Multiple alleles means that there are **more than two possible alleles** for that trait, although each person inherits **only two of the three possible alleles** (one from their mother and one from their father).

In human blood types, the alleles I^A and I^B are codominant. Both I^A and I^B are **dominant** over the **recessive allele i.**

TABLE 12-3

BLOOD TYPES AND THEIR GENOTYPES

PHENOTYPE (BLOOD TYPE)	ANTIGENS PRESENT	POSSIBLE GENOTYPES
A	antigen A only	I^AI^A or I^Ai
B	antigen B only	I^BI^B or I^Bi
AB	both antigens A and B	I^AI^B
O	neither antigen A nor antigen B is present	ii

1. Howard Hughes was an inventor, multimillionaire businessman, and adventurer, whose life was the subject of the 2004 film *Aviator*. When he died, he left no legitimate heirs. Soon, however, a long succession of people claiming to be his children began to appear.

    A young man claiming to be Howard Hughes's child sued for a share of the estate. The judge ordered blood tests to determine the validity of the claim. Howard Hughes had blood **type AB,** the **mother** of the young man had blood **type A,** and the **young man** himself had **type O** blood. If you were the judge, how would you rule? **Explain** your answer.

2. Is it possible for a **type A** person married to a **type B** person to have **type O** children? **Explain** your answer.

Check your answers with your instructor before you continue.

SELF TEST

1. The ability to taste PTC is due to a **dominant allele (T).** A woman nontaster married a man who was a taster. They had three children. Their two sons were tasters, but their daughter was a nontaster. Both parents of the woman and both parents of her husband were tasters. **What are the genotypes** of all the individuals mentioned?

    The woman _____   Her husband _____

    The two sons _____   The daughter _____

    The woman's parents _____ and _____

    The husband's parents _____ and _____

2. Your sister died from Tay-Sachs disease, inherited as a **recessive** allele **(t).** You're married and planning to start your family. You're worried about the disease and decide to have genetic testing to see if you or your spouse is a carrier of the Tay-Sachs allele. The test results show that you're a carrier of the allele, but your spouse isn't.

    What is the probability that you and your spouse will have a child with Tay-Sachs disease? **Show your work.**

3. Red-green color blindness is inherited through an **X-linked, recessive** allele **(b).** Two parents, Fred and Ginger, have normal vision. They have two daughters, Takiyah and Kelly, who also have normal vision, and a color-blind son, David.

    Daughter Kelly has a color-blind son, Kevin. Daughter Takiyah has five sons, all with normal vision. What are the genotypes of all the individuals? **Show all your work!**

    Fred _____   Ginger _____   David _____

    Takiyah _____   Kelly _____   Kevin _____

    Takiyah's five sons _____

    If Kelly marries a man with normal vision, what is the probability that she'll have a color-blind son? _____ a color-blind daughter? _____

4. Ralph has normal blood clotting, but he has two brothers and a sister who have hemophilia (an **X-linked, recessive** disorder). What are the most probable genotypes of Ralph's parents? **Explain** your answer.

5. Albinism in humans is expressed as the **absence** of pigment from the skin, hair, and eyes. Using the information in **Figure 12-6,** determine whether albinism is inherited as a **dominant or as a recessive** trait. **Affected** individuals are represented by **shaded** squares and circles.

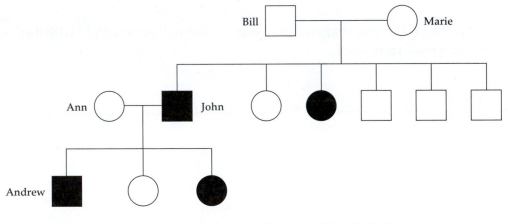

FIGURE 12-6. Pedigree Showing Albino Individuals

a. Underneath **each circle and square** in the pedigree, **enter the genotype** of that individual.

b. **(Circle one answer).** Albinism is a trait that is probably inherited through a **dominant / recessive allele.**

Support your answer with information from the pedigree.

6.   Neil and Alice are concerned because their newborn daughter doesn't appear to resemble either of them. They suspect there was a mix-up at the hospital. They check the blood type of the baby and find it is **type O.** Because Neil has **type A** blood and Alice has **type B,** they conclude that a mistake has been made. Are they correct? **Explain** your answer.

EXERCISE

13

Evolution

Objectives

After completing this exercise, you should be able to:

- describe the contributions of Charles Lyell and Charles Darwin to the formation of the theory of natural selection
- explain the process of radiometric dating and its use in determining the age of rock strata
- explain the use of index fossils to date sedimentary rock strata
- discuss several ways in which the reconstruction of fossil remains can provide information about an organism's physical characteristics, habitat, and geographic distribution
- explain the process of natural selection in the context of the evolution of antibiotic-resistant bacteria

CONTENT FOCUS

In the 1800s and earlier, people had very different ideas about life on earth than we have today. The prevailing thoughts in Europe were that species never changed and the earth was about 6,000 years old.

The beginnings of a change in scientific thought were made possible by the discoveries of **Charles Lyell,** a geologist who traveled through Europe and made observations that supported the hypothesis that the surface of the earth is gradually altered by natural processes. For example, rain and wind weather away mountains, volcanic activity produces new land, and older layers of sediments are lifted up by movements of the earth's crust. These changes occur slowly, so Lyell concluded that the age of the earth was much more than 6,000 years.

Charles Darwin was influenced by Lyell's discoveries. He hypothesized that if the earth's features could change, then perhaps living species could also change. He traveled the world for five years on a sailing vessel named the *Beagle*. Every place he landed, he looked for evidence to support his hypothesis. Eventually he accumulated so much evidence that slow changes had occurred that he proposed a new theory of evolution that occurred by a process he called **natural selection.**

ACTIVITY 1
DETERMINING THE AGE OF ROCK LAYERS AND FOSSILS

How are fossils formed? In some situations, dissolved minerals gradually seep in and replace parts of the body exactly, leaving a shell or skeleton made of stone. For example, the fossilized fish skeleton in **Figure 13-1** is preserved in detail. Other fossils form as molds or impressions in soft sand or mud, and those sediments are later compressed into rock.

FIGURE 13-1.   Fossilized Fish Skeleton

Lyell and Darwin accumulated compelling evidence that the earth was much older than previously suspected. For example, Darwin compared fossils of extinct species with modern species living in the same geographic area. In many cases, they were remarkably similar. As Lyell examined deeper (older) rock layers, he found that there were no fossils of modern animals. As he examined rock layers progressively closer to the surface, however, he found a sequence in the types of fossil remains present. In the case of animals, for example, Lyell and Darwin found evidence of early fishes, amphibians, reptiles, birds, and finally modern mammals (including humans).

How is it possible to find out how much time had passed between ancestors and descendants? Complete the following activity that demonstrates two methods for determining the age of rock layers and the fossil remains found in those layers.

One method of documenting changes in fossil remains is by the relative positions of rock layers (called **strata**). Consider the visible rock layers in **Figure 13-2.** The rock layer closest to the surface is the most recently formed and therefore, the "youngest" of the strata. Each rock layer is formed on top of the previous surface. Therefore, the older layers are deeper in the rock formation.

FIGURE 13-2. Exposed Rock Layers

In the early 1900s, it was discovered that the chemical elements found in rock layers can exist in multiple forms called **isotopes.** Isotopes can be converted into other elements over time, a process known as **radioactive decay.** For example, lead is the **breakdown product** of the decay of an isotope of uranium 238 (^{238}U).

You can date a rock layer by comparing the amount of the radioactive element to its breakdown products in a specific rock layer. The process is known as **radiometric dating.** Scientists use the half-life of an element to make their calculations.

The **half-life** is the amount of **time it takes for one-half of an isotope sample to decay** into a different element. The half-life of ^{238}U is 4.5 billion years. Other radioactive isotopes decay in a much shorter time. By comparison, the half-life of ^{14}C (carbon 14) is approximately 5,700 years. Radiometric dating has calculated the age of the earth at approximately **4.6 billion** years.

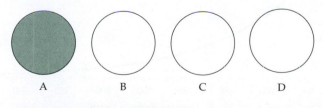

FIGURE 13-3. Half-Life of a Radioactive Isotope

1. Which circle in **Figure 13-3** represents the amount of the original isotope before decay began? _____

2. Color in the area of Circles B, C, and D that represents the **amount of ^{238}U remaining** in the rock layer as each half-life passes.

3. If **Figure 13-3** represented the half-life of ^{238}U, which circle that you colored would represent a rock layer with the greatest concentration of lead? _____

4. Which circle that you colored shows the amount of the original isotope remaining after **two** half-life periods have expired? _____

5. If **Figure 13-3** represented the half-life of ^{14}C, how many years would have passed to reach letter D? _____ years

   You're a paleontologist examining igneous rock layers. You're waiting for the test results from the samples you took from the layer you're trying to date. The graph in **Figure 13-4** shows the test results returned from the laboratory.

6. Based on the graph in **Figure 13-4,** the half-life of the isotope shown is ____ million years.

7. If you took a rock sample that contained 30 grams of the original isotope, the rocks that sample came from were about _____ years old.

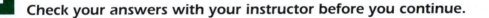

Check your answers with your instructor before you continue.

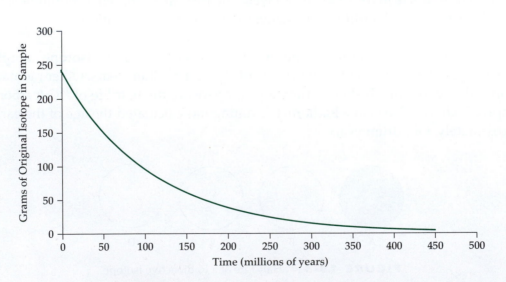

FIGURE 13-4.   Rate of Decay of Rock Sample

The method of radiometric dating you just completed works best for volcanic rocks because they contain radioactive elements, but it doesn't work as well for sedimentary rocks. However, sedimentary rocks can also be dated by the presence of specific fossils found in particular layers. Within each sedimentary layer are the fossil remains from organisms that existed only during limited periods of geologic time (called **index fossils**). Index fossils are representative of only one specific time period and so they can be used as indicators of the age of the rock strata in which they are found.

When sedimentary rock layers are undisturbed, the fossil record is clearly shown in the sequence and age of the layers in the formation. However, in many places around the world, rock layers have shifted or been uplifted, so that the sequence of sediment deposition is difficult to determine. However, these layers can be dated and sequenced by comparing the index fossils with those in other, undisturbed locations.

In a simple demonstration of this method of comparison, correlate the rock layers in **Figure 13-5. The letters in the layers represent specific index fossils found only in that layer.**

8. Draw lines that connect the matching index fossil layers in **Figures 13-5a** and **13-5b**.

9. Which is the oldest rock layer in **Figure 13-5?** _____

    Which is the youngest? _____

(a)   (b)

FIGURE 13-5.   Index Fossils Embedded in Sedimentary Rock Layers

10. Which rock layers in **Figure 13-5a** have matching layers in **Figure 13-5b**?

11. What are some possible explanations for the fact that some rock layers are thicker than others?

12. Some of the index fossils in **Figure 13-5a** aren't shown on the diagram in **Figure 13-5b**. What is the most likely reason for this occurrence?

13. **Challenge Question!** What environmental factors in the location where the rock layers in **Figure 13-5a** formed might account for the fact that there are many more layers present there than in the location shown in **Figure 13-5b**?

Check your answers with your instructor before you continue.

ACTIVITY 2   RECONSTRUCTING FOSSIL EVIDENCE

Through the fossil record, we can gain information on plants and animals from past eras—even if they are no longer alive in modern times. Fossil evidence from these early remains has allowed scientists to establish a lineage of plants and animals through time.

However, the fossil record is incomplete and takes knowledge to interpret. In this activity, you'll gain some insight into the methods scientists use to reconstruct fossil remains. Through observation and interpretation, you can discover details of how the organisms looked, their geographic distribution, and even gain some information about their lifestyles.

> ### Note:
>
> **During this activity, it's less important to "get the right answer" than it is to expand your understanding of how scientists use fossil evidence to answer questions about the evolution of life on earth.**

1. Work in groups. From the supply area get **a set of three envelopes marked for your group.**

    In the envelopes are fossilized dinosaur bones collected during a recent dig in the western United States. The rock strata at the excavation site were dated to the Upper Cretaceous Period (from 76 to 65 million years ago).

    The bones you'll be analyzing were collected on several successive days of the dig. **You'll examine each day's collection of bones in sequence** and gradually accumulate information about the types of dinosaurs the bones belonged to and their habitat.

2. Remove the bones from the envelope marked "**Day 1.**"

    Form some hypotheses:

    a. What parts of the dinosaur do you think your bones came from?

    b. Do all the bones appear to belong to the same type of dinosaur? _____

3. If you've hypothesized that you have the bones of more than one species, divide the bones you think belong to each type of dinosaur into separate piles.

> ### Hint
>
> **Keep in mind that you may have to revise your hypotheses about the collected bones when you get additional information.**

4. Remove the bones from the envelope marked "**Day 2.**" Add them to your collection from **Day 1.** Observe the bones and try to determine if any of them belong together or are part of the same body structures.

    Don't attempt to build the model! With the bones lying **flat on the table top,** try to arrange the bones into an approximation of how they might have been positioned in the living animal.

    Do all the bones in your collection appear to belong to the same dinosaur? Explain your answer.

5. Remove the bones from the envelope marked "**Day 3.**" Add the bones to the others that are grouped on the table top.

    What new information was added?

6. Do any of the bones you discovered resemble bones from your own skeleton? If so, which ones?

Note:

Your collaboration will be more effective if you view your bones together with those of other groups. However, leave your Day 1 to Day 3 envelopes on your table when you carry your bones away. When the activity is completed, you'll need to return the bones to their proper envelopes (by matching the bones to the numbers on each envelope).

7. At the present time, you have partial skeletons of one or more dinosaurs. You know that other colleagues have been on digs in rock strata from the same geological period and have uncovered numerous bones. Collaborate with **one or more groups** and compare your bones with their discoveries.

   Has the ability to add bones to your skeletal reconstruction(s) led you to change any of your hypotheses? If so, what changes did you make? Explain your answer.

8. Do you think your dinosaur(s) fed on meat or plants? What **fossil evidence** (body form or structures) supports your hypothesis?

9. What form of locomotion do you think your dinosaur(s) used? What **fossil evidence** supports this hypothesis?

10. Other types of fossils found in the same rock layer with your reconstructed skeleton might give you some clues about the ecological conditions at the time that dinosaur was alive. For example, if you found fossilized dung containing fish bones, what type of habitat do you think your species lived in? Why?

11. You discover from your research that similar fossils have been found elsewhere in the United States, Canada, China, and North Africa. What does this tell you about the environmental conditions in these locations during the Upper Cretaceous? Explain your answer.

12. What parts of the scientific method did you use in your experience with the skeletal reconstruction? Explain your answer.

Check your answers with your instructor before you continue.

ACTIVITY 3   NATURAL SELECTION

Evolution can be defined as a shift in the percentage (**frequency**) of certain **alleles** (forms of a gene) in a population. There are several mechanisms by which some alleles can become more common than others—one of these mechanisms is natural selection.

For example, the frequency of a single sickle cell allele is common in human populations that live in regions of Africa where malaria is prevalent, despite the fact that inheritance of two sickle cell alleles leads to serious disease and frequently death at a young age. The **shift in the frequency of this allele** has occurred because its presence confers a **survival advantage** to individuals who have one copy of the allele. What advantage can a disease allele possibly provide? In this case, it provides protection from the most serious effects of malaria infections and therefore results in differential survival of people in the population with various genotypes.

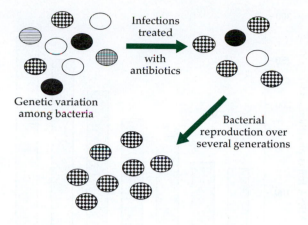

FIGURE 13-6. Evolution of Disease-Causing Bacteria

Individuals in any population (plants, animals, microorganisms) differ from one another, and many of these differences are genetically based. This principle is referred to as **genetic variability.** Individuals who are born with **inherited** traits that **improve their ability to survive** are more likely to become adults and the parents of the next generation. They pass these inherited "survival" traits on to many of their offspring, and over time, more members of the population will share those beneficial traits. **Natural selection** has occurred.

Because natural selection involves genes passing from generation to generation, the process is easier to observe in species that have a short generation time (in comparison to human generations). An example of the evolution of bacteria (which reproduce quickly) after treatment with antibiotics is shown in **Figure 13-6.**

1. Why aren't all the original genetic variants in the bacterial population present in the same proportion as they were before treatment with the antibiotic?

2. What caused the shift in allele frequencies from the initial population to that observed after several generations?

3. What term describes the process of evolution illustrated in **Figure 13-6?**

In the previous example, the population has shifted from being susceptible to antibiotics to a population with many more resistant individuals. This pattern of natural selection is called **directional selection,** because the allele frequency has shifted in a specific direction.

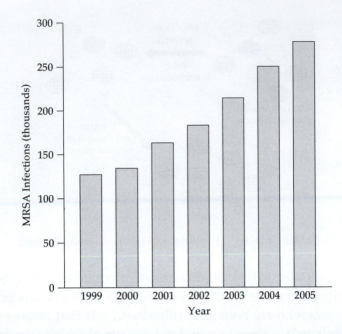

FIGURE 13-7.   Patients Hospitalized with MRSA Infections 1999–2005
Source: Centers for Disease Control and Prevention, 2007.

In recent years, directional selection has been observed in a common type of bacteria that causes a sore throat and other types of more serious infections. The species is *Staphylococcus aureus* (the cause of a "staph" infection). A strain of staph bacteria has become resistant to the antibiotics that were previously used to treat these infections.

The resistant variety is known as MRSA (methicillin-resistant *Staphylococcus aureus*). Most MRSA infections are contracted during hospital and dialysis-center visits, among nursing home residents, and at child care centers.

4. Summarize the information shown in **Figure 13-7**.

5. In your own words, explain the information shown in **Figure 13-7**. In your explanation, use the following terms: **evolution, directional selection, antibiotic resistance, survival advantage,** and **allele frequency**.

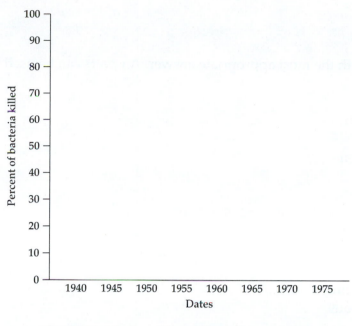

FIGURE 13-8.   Effect of Penicillin on Bacteria

6. Penicillin was first discovered in 1929 and came into common use in 1940 at the start of the U.S. involvement in World War II. Draw a line on the blank graph in **Figure 13-8** that shows your hypothesis about the effectiveness of the drug in curing bacterial infections over the duration of the war (1940–1946).

    Extend the line on the graph showing your hypothesis about the effectiveness of penicillin in killing bacteria from 1946 through 1975.

7. **Explain the logic** you used to draw your line on the graph in **Figure 13-8**.

Check your answers with your instructor before you continue.

SELF TEST

Fill in the blanks with the most appropriate answer. Answers can be used **only once.**

a.   Charles Lyell
b.   Charles Darwin
c.   index fossil
d.   natural selection
e.   evolution
f.   paleontologist
g.   half-life
h.   sedimentary
i.   isotope
j.   fossil record
k.   radiometric dating
l.   Upper Cretaceous
m. ^{14}C dating

1. _____ This method could be used to determine the age of dinosaur fossils.

2. _____ A geologist who provided evidence that the earth was much more than 6,000 years old.

3. _____ The notation ^{235}U is used to designate a(n) ___ of an element.

4. _____ A change in the frequency of a specific allele in a population over time.

5. _____ Proposed the theory of natural selection.

6. _____ Rock layers formed by the accumulation and consolidation of sand or mud.

7. _____ Organisms that inherit alleles that help them survive are more likely to reproduce and pass those traits on to their offspring.

8. _____ Remains of plants, animals, and microorganisms from earlier eras.

9. _____ A geologic era that ended approximately 65 million years ago.

10. A pharmaceutical company has developed a new and more effective antibiotic. It predicts that it will make a big profit on this new drug and that the money will keep rolling in over the course of several decades. Based on your knowledge of **directional selection,** do you agree with this financial forecast? Explain your answer.

11. You're a paleontologist trying to date the rock layers at your dig. You remove a sample and discover that the rock contains 5 mg of ^{235}U. The half-life of ^{235}U is 700 million years. Data from samples show that when the same rock was newly formed, it contained 20 mg of ^{235}U. How old is the rock stratum at your dig? Show your work.

12. Explain how genetic variation is related to evolution.

Functions of Tissues and Organs

Objectives

After completing this exercise, you should be able to:

- list and explain the function of several features present in each of the three layers of the skin
- explain how each feature in the epidermis, dermis, and hypodermis demonstrates the relationship between structure and function
- describe at least four examples of how deposits of fats and oils are useful to plants, animals, and microorganisms
- discuss the relationship between the density of touch receptors in various body locations and touch sensitivity
- explain the effects of muscle fatigue on muscle action

CONTENT FOCUS

Within our bodies, we have approximately 100 trillion cells. As you observed in previous exercises, not all cells have the same **structure.** In the bodies of multicellular animals such as humans, not all cells have the same **function.**

A **tissue** is a **group of similar cells that perform a specific function.** Cells are organized into **four major tissue types:**

Epithelial Tissue	Lines all inner and outer body surfaces; covers organs and body cavities
Muscle Tissue	Contracts to produce movements
Connective Tissue	Joins and supports other tissues
Nervous Tissue	Senses stimuli; transmits signals around the body

Tissues group together to form the organs of the body. All of these tissue types are present in the **skin,** the body's largest organ.

ACTIVITY 1   SKIN: THE OUTER PROTECTIVE LAYER

1. Get a dropper bottle of water. While holding your hand level in front of you, palm-side down, place a drop of water on the back of your hand. What happens to the water droplet?

2. Did the liquid penetrate the skin easily? _____

   Is this one of the **protective functions** of the skin? _____ **Explain** your answer.

3. The **waterproofing** quality of the skin is due to the presence of the **protein keratin** in epithelial cells. The **keratinized** cells of the epidermis also protect the under-lying tissues from **mechanical injuries and abrasions.** On parts of the body where **abrasion is most common,** the epidermal layer tends to be **thicker.**

4. The skin also provides protection from exposure to **ultraviolet (UV) radiation.** Pigment-producing cells in the epidermis manufacture the **protein melanin** that blocks penetration of UV rays and protects underlying cells from damage.

   Most people have about the same number of melanin-producing cells, called **melanocytes. Dark-skinned** individuals, however, produce **more and darker melanin** than fair-skinned individuals do. Melanin production is also **stimulated by exposure to sunlight.**

5. Explain how the following **factors work together** to form a protective barrier in the skin: **ultraviolet radiation, melanocytes, melanin,** and a **suntan.**

6. Problems with melanin production can result in different disorders.

    Albinos have a normal number of melanin-producing cells, but the cells don't synthesize melanin. For this reason, albinos have no pigment in their skin, eyes, and hair.

    In the condition **vitiligo,** melanocytes die, causing patches of skin to lose their coloration. These light spots in the skin are often surrounded by skin with normal pigmentation.

✔ Comprehension Check

1. To raise some money for your tuition, you take a summer job working construction. After several weeks, you notice that **calluses** have developed on your hands. What is a **callus? Why did the calluses form?**

2. The incidence of **skin cancer** is increasing in our population. **(Circle one answer.)** You have better protection from exposure to the damaging effects of UV rays if you have **dark / fair** skin.

Check your answers with your instructor before you continue.

ACTIVITY 2   THE EPIDERMIS

By looking at **Figure 14-1,** you can see that the skin is made up of several layers: the **epidermis** (the outermost layer), the **dermis** (the middle layer), and the **hypodermis** (the innermost layer).

The epidermis consists of many layers of **epithelial cells** but comprises only a small portion of the total thickness of the skin. The protective layer of keratinized cells at the **surface** of the epidermis is **repaired by rapidly dividing cells** at the **base** of the epidermis. Newly formed cells are gradually pushed upward to replace lost and damaged cells in the outer layer.

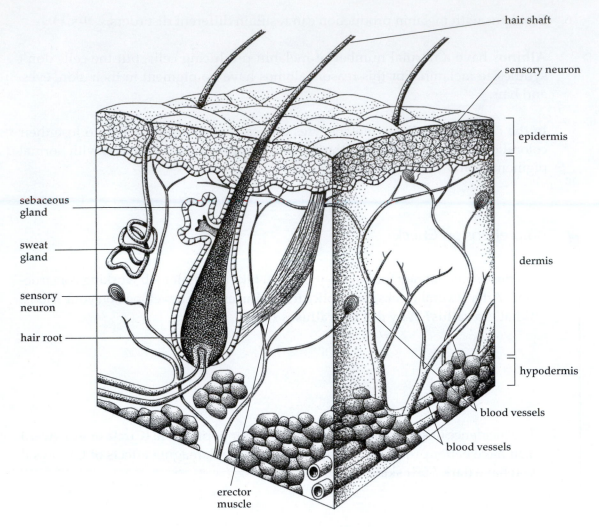

FIGURE 14-1. Layers of the Skin

✔ Comprehension Check

1. In some individuals, excessive shedding of cells occurs in the scalp area. These discarded scalp cells are referred to as _____.

2. On **Figure 14-1, draw an arrow** showing the direction of cell replacement in the epidermis.

3. **(Circle one answer.)** Actively dividing cells can be found in the **top / middle / bottom** layer of the epidermis.

4. On **Figure 14-1, draw an arrow** on the layer of the epidermis in which cells are dividing.

| 5.   Using a colored pencil or highlighter, **color the blood vessels** in **Figure 14-1.**

Check your answers with your instructor before you continue.

ACTIVITY 3   THE DERMIS

1. By looking at **Figure 14-1, circle the correct answer** in each of the following statements:

    The dermis is located **above / below** the epidermis.

    The dermis is **thicker / thinner** than the epidermis.

    Hair roots are located in the **dermal / epidermal** layer of the skin.

2. Take a look at the epidermal and dermal layers in **Figure 14-1.**

    Are any **blood vessels** present in the **epidermis?** _____

    If you cut **just** your epidermis, will the cut bleed? _____ **Explain** your answer.

3. Both **sweat glands** and **sebaceous glands** are found in the dermis. Sweat glands function in body cooling and release sweat through **ducts** that extend to the surface of the epidermis (see **Figure 14-1**).

    Sebaceous glands produce oil that lubricates the hair and skin. Sebaceous glands have **no ducts** to release their oil to the skin surface, so they make use of the space created by the hair shaft to transport the oil (see **Figure 14-1**).

✔ Comprehension Check

1. The sale of **hand lotion** is a multimillion-dollar market in the United States. Using your knowledge of how the epidermis is lubricated, explain why hand lotion is such a big-selling item. In your answer, include the following terms: **sebaceous gland, duct, hair shaft,** and **oil.**

2. Explain why the skin of a person's scalp is often more oily than the skin of his/ her arm.

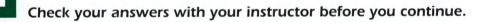

Check your answers with your instructor before you continue.

ACTIVITY 4   HAIR TODAY, GONE TOMORROW

Millions of hairs are scattered over the body. The average scalp has about 100,000 hairs, and a man's beard has another 30,000.

1. Refer back to **Figure 14-1.** You'll see that a hair consists of two parts, the **shaft** and the **root.** The shaft portion of the hair begins in the **dermis** and extends **outside** the surface of the skin.

    The **root** is found **only within** the dermis and is surrounded by several layers of cells. Collectively, this structure is called a **hair follicle.**

2. Sit with your **eyes closed** while your partner runs his/her hand over the top of your head, **gently** touching the **hairs.**

    Can you tell when you are being touched? _____

    What **section** of the hair was just touched? _____

    A knot of **sensory nerve endings** is wrapped around the **root of each hair.**

    Moving a hair activates these nerve endings, so that your hairs function as sensitive touch receptors.

3. Refer back to **Figure 14-1.** You can see that each hair is connected to a **tiny erector muscle** that **automatically** raises the hair when you're cold.

    In animals with fur coats, raised body hairs **hold a layer of warm air** next to the skin, insulating them from the cold. Humans have too few body hairs for this response to help keep us warm, but we are like other animals in our response to changing temperatures.

✔ Comprehension Check

1. Which of the following skin features can be found in the **dermis? (Circle ALL correct answers.)**

    a. melanin
    b. blood vessels
    c. nerve endings
    d. sebaceous glands
    e. erector muscles

    f. hair follicle
    g. hair shaft
    h. highly keratinized cells
    i. sweat glands

2. If you get a paper cut that penetrates to the **dermis,** will the cut bleed? _____

Check your answers with your instructor before you continue.

ACTIVITY 5   BELOW THE SKIN (THE HYPODERMIS)

1. Beneath the skin is a layer of **connective tissue** with many fat cells. (See **Figure 14-1.**)
    This layer is called the **hypodermis** (**"hypo"** means **under**) or **subcutaneous** layer
    (**"sub"** means **under** and **"cutaneous"** refers to the skin).

    Perform the following experiment using a **"blubber mitten"** to discover one of the
    functions of the subcutaneous fat layer.

2. Work **individually.** Get the following supplies: **one blubber mitten and one empty
    mitten.**

    The two mittens are designed to simulate how the subcutaneous layer of the skin
    would function **with and without fat cells.**

    In this experiment, you'll place **one mitten on each hand** and **immerse both hands**
    in an **ice-water bath.**

3. **Before proceeding,** form a hypothesis about the **effect of water temperature** on
    each of your hands. **Record** your hypothesis below.

HYPOTHESIS FOR BLUBBER MITTEN EXPERIMENT

4. Go to the ice-water bath and **perform your experiment. Record your results**
    below.

RESULTS OF BLUBBER MITTEN EXPERIMENT	
WITHOUT FAT	WITH FAT

5. From the results of your experiment, what **conclusions** can you draw about the
    effect of fatty tissue in the subcutaneous layer of the skin?

Comprehension Check

1. Which of the two mittens was the **control** in your experiment?

2. Why was it **necessary** to use a **control mitten** (instead of putting your **bare hand** in the ice-bath)?

3. **Based on the results** of your experiment, give **one function** of subcutaneous fat tissue. _____

4. Again, **referring to the results** of your experiment, suggest **two reasons** why some animals eat a lot in the fall season?

   a.

   b.

5. You're probably familiar with the fact that when you **mix oil and water,** the oil will float to the surface. Do you think that significant **fat** deposits, such as those found in marine mammals, contribute to their **buoyancy? Explain** your answer.

6. On an ocean field trip with your marine biology class, you make a collection of **phytoplankton (tiny marine plants).** When you look at these plants through the microscope, you notice that many of them have **oil droplets** inside their cells.

Suggest **two ways** in which oil droplets might be helpful to the phytoplankton.

a.

b.

Check your answers with your instructor before you continue.

ACTIVITY 6   TOUCH SENSITIVITY

The nervous system acts as an interface between the body and the outside environment. Information about the outside world is acquired through activation of sensory neurons that act as receptors in the dermis and epidermis of the skin.

Sensory neurons process and transmit incoming information to the **central nervous system (the brain and spinal cord).** Neurons vary in function and appearance, but all share a similar three-part structure: the **cell body,** the **axon,** and the **dendrites** (see **Figure 14-2**). **The arrows on Figure 14-2** indicate the **direction** of impulse transmission (from the dendrites to the cell body, and then to the axon).

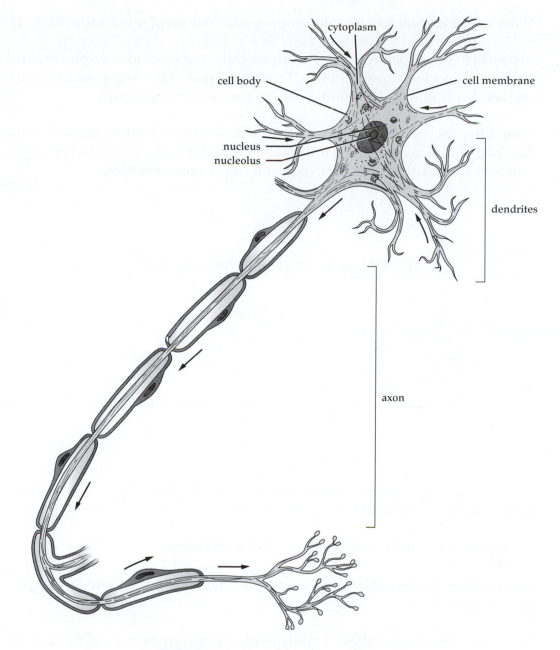

cytoplasm

cell body

cell membrane

nucleus
nucleolus

dendrites

axon

FIGURE 14-2. Structure of a Typical Neuron

Every neuron has a **cell body,** the region of the cell that contains the nucleus and a variety of other organelles necessary for cell metabolism. Outside stimuli are received by **dendrites** in the skin and other sense organs. Incoming signals from these dendrites are transmitted toward the cell body and continue into the **axon.** Axons relay outgoing messages from one neuron to other neurons or to various tissues and organs in the body.

Dendrites in the skin are specialized to receive different types of sensory input, such as **touch, temperature, pressure, pain, vibration,** and **proprioception** (the sense that tells you the current position of your body and limbs). In the following experiment, you'll be making observations about the distribution of touch receptors in the skin.

1. Work **with a partner.** Get the following supplies: **one set of touch calipers.**

    The set of touch calipers consists of a series of corks. Each cork has two pins inserted so that the blunt ends are protruding. The pins are spaced at varying distances from one another **(1 mm, 2 mm, 4 mm, 8 mm, 16 mm,** and **32 mm apart).**

2. Form a hypothesis about the relative sensitivity of various body locations. List the **four body locations** mentioned in **Table 14-1,** in order from the area that you think will be **least sensitive** to the area you think will be **most sensitive.**

TOUCH SENSITIVITY HYPOTHESIS

3. Perform the following set of tests on your partner, and then reverse roles and have your partner use the same tests to check your touch response sensitivity.

    Your partner should be seated, with his/her **eyes closed.**

    For each touch, the subject will report whether he/she feels one or two pins touching the skin.

4. Begin with the **32-mm caliper.** Lightly touch the skin on the back of the test subject's hand.

    Repeat the process with the remaining calipers in the set **(16 mm, 8 mm, 4 mm, 2 mm,** and **1 mm) in any order you wish.**

    For each test, record whether your partner is able to feel two pins touching the skin.

    If the test subject felt **two pins,** record **yes.** If the test subject felt **one pin only,** record **no.**

    Record these results in **Table 14-1.**

5. Repeat the same procedures at each of the following locations:

    inside of the forearm

    tip of the nose

    tip of the forefinger

6. **Reverse roles** and have your partner repeat the same tests to check your touch
    response sensitivity. Record the results in **Table 14-1.**

T A B L E   1 4 - 1
TOUCH SENSITIVITY IN VARIOUS BODY LOCATIONS

TEST SUBJECT 1						
BODY LOCATION	32 mm	16 mm	8 mm	4 mm	2 mm	1 mm
Back of hand						
Inside of forearm						
Tip of nose						
Tip of forefinger						
TEST SUBJECT 2						
Back of hand						
Inside of forearm						
Tip of nose						
Tip of forefinger						

The **greater the number of touch receptors** in a body location, the **more sensitive**
that location will be to touch stimuli. The reason for the increased sensitivity lies in the
distribution of the receptors. If the pin points touch **two receptors that are adjacent** to
each other, the pins will **activate both receptors together** and it'll feel as though you
have been touched by **only one pin.**

If the pins touch two receptors that are **not adjacent,** however, each will **fire separately** and you'll feel the touch of **both pins.** The ability to feel the two pins is called **"two-point discrimination."**

In locations with a high density of touch receptors, therefore, the skin is much more sensitive to touch and much better at providing information about the number of touches (one pin or two pins). This type of touch sensitivity allows a surgeon or mechanic to perform delicate procedures working by touch alone.

The smallest pin distance when the test subject had effective two-point discrimination indicates the body location with the highest density of touch receptors.

The higher the number of touch receptors in a specific location, the greater the probability the pins can stimulate two nonadjacent touch receptors.

Comprehension Check

1. Based on the results in **Table 14-1,** rank all areas tested in terms of their relative sensitivities to touch. List the body locations from the area of **least sensitivity** to the area of **highest sensitivity.**

   Subject 1:

   Subject 2:

2. Which of the body locations tested has the **greatest density** of touch receptors? **Explain** your answer.

3. Would you expect a decrease in touch sensitivity for each of the following conditions? **Explain each answer.**

   a. calluses on the palms of your hands:

   b. thick subcutaneous fat layer:

4. People who are visually impaired are able to read books and other printed materials by using the Braille system of writing. In the Braille system, words are represented by patterns of raised dots. Braille readers touch the dot patterns with their fingertips to read the text.

   How does the density of touch receptors in the fingertips make the Braille system possible?

Check your answers with your instructor before you continue.

ACTIVITY 7   MUSCLE FATIGUE

During exercise, contracting muscles become progressively weaker until the muscle cells no longer respond to stimulation. This process is known as **muscle fatigue.** Muscle fatigue can be explained by several factors that occur simultaneously as your body works. These include:

- lack of ATP to meet energy needs
- insufficient oxygen for cell respiration
- depletion of energy reserves in the muscle cells

1. Work in groups. During this experiment, you'll be supporting a **heavy** book on your open palm with your arm completely extended. Which arm can support the book longer? **Form a hypothesis** about the ability of the muscles in your right and left arms to support the book and **record** your hypothesis below.

HYPOTHESIS FOR MUSCLE FATIGUE EXPERIMENT

Check your hypothesis with your instructor before you continue.

2. For your muscle fatigue experiment, one member of the group will act as a **time-keeper,** the second will **record** the experimental results, and the third will be the **test subject.**

Note:

The person holding the book shouldn't know how much time has expired until the entire experiment is completed.

---

3. Get an **extremely heavy book.**

   a. Place the book on your open hand.

   b. With the book on your hand, **extend your arm fully with the palm up. Don't bend your elbow.**

   Your arm should **not** be braced against your body.

   c. Record the length of time you can hold your arm extended. **Record** the results in **seconds** under **Trial 1** in **Table 14-2.**

   d. Rest your arm for **5 seconds** and repeat the experiment for **Trial 2.**

   e. Rest your arm for **5 seconds** and repeat the experiment for **Trial 3.**

   f. Rest your arm for **5 seconds** and repeat the experiment for **Trial 4.**

TABLE 14-2 RESULTS OF MUSCLE FATIGUE EXPERIMENT		
TRIAL NUMBER	TIME ARM HELD EXTENDED (sec) RIGHT ARM	TIME ARM HELD EXTENDED (sec) LEFT ARM
1		
2		
3		
4		

4. **Graph** your experimental results in **Figure 14-3.** Plot **time** on the **Y-axis.**

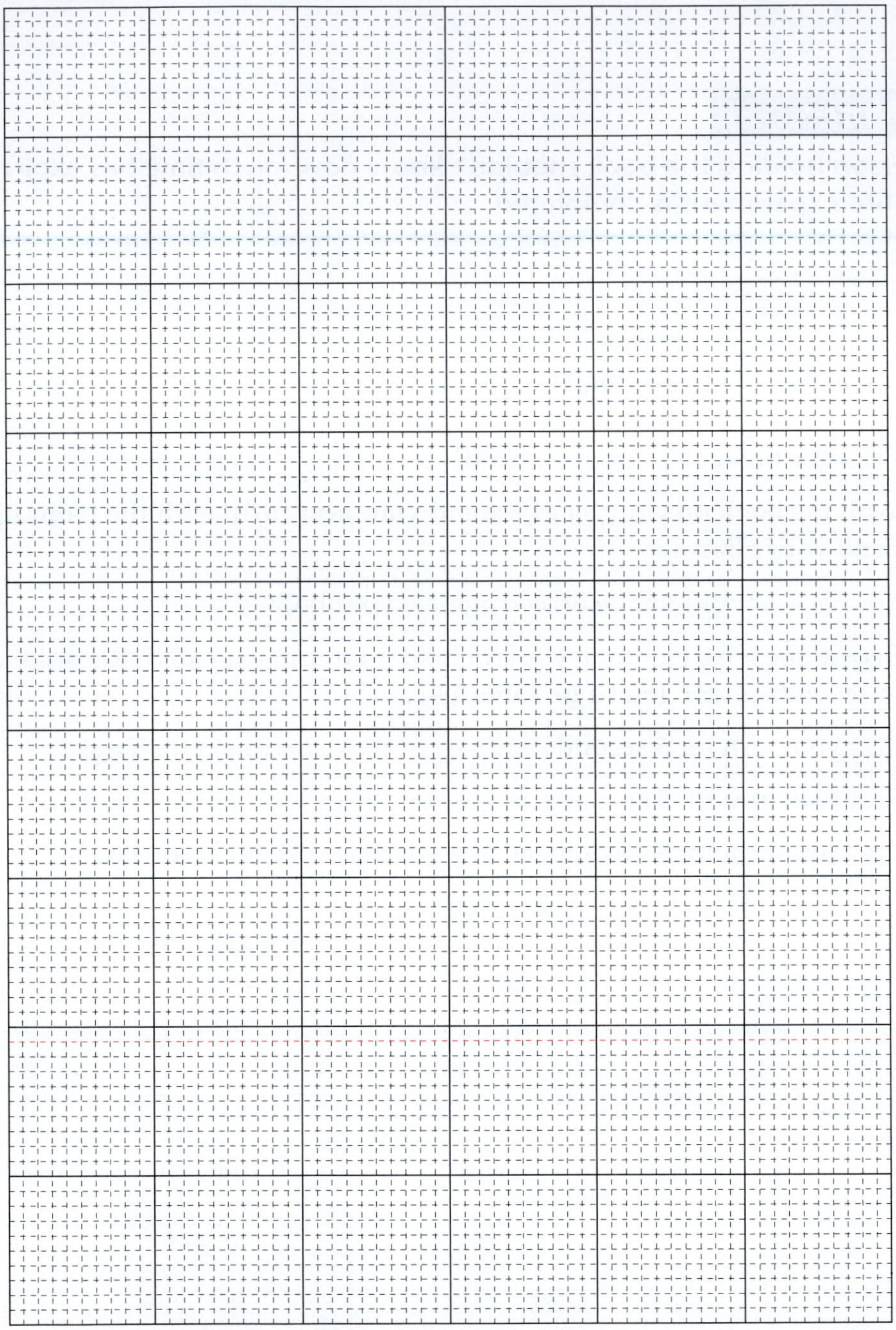

FIGURE 14-3.   Comparison of Muscle Fatigue in Right and Left Arms

5. Did the experimental results **support** your hypothesis? _____

    Explain your answer, mentioning **facts** collected during your experiment.

6. Did you see evidence of **muscle fatigue? Explain** your answer.

7. To have full confidence that your conclusions about muscle fatigue are accurate, **what changes would you recommend** for the experimental design?

Check your answers with your instructor before you continue.

SELF TEST

Fill in the blank with the choice that is **most appropriate** to describe the function of each skin structure. Answers can be used **only once.**

a. keratin
b. melanin
c. erector muscle
d. epithelial cell
e. epidermis

f. hair follicle
g. blood vessels
h. subcutaneous layer
i. sebaceous gland
j. dermis

1. _____ Produces oil to lubricate the hair and skin.

2. _____ Layer of skin containing many blood vessels and nerves.

3. _____ Type of protein deposits in the skin that form fingerprints.

4. _____ Type of dead cell that flakes off as dandruff.

5. _____ When the hairs on the back of a dog stand up, this tissue is responsible.

6. _____ Location of fatty tissue that insulates and protects the body.

7. _____ Accumulation of this protein helps protect your skin from the sun's rays.

8. _____ The outermost layer of the skin.

9. You're determined to quit smoking, and so have just started using a nicotine patch on your arm. In which layer of skin does the nicotine first enter the bloodstream? **Explain** your answer.

10. In reference to the nicotine patch, even though it wouldn't be as visible, why would it be a bad idea to place the patch on the sole of your foot? (Answer in terms of the ability of the patch to **function.**)

11. You and a friend are at the gym working out. Your friend is upset because she can't do as many repetitions of a bench press in her third set as she did in her first set. What would you tell her?

12. **Challenge Question!** Burns are classified according to how many skin layers are destroyed. A **first-degree burn** affects only the outer layers of the epidermis. A **second-degree burn** destroys the epidermis and penetrates the dermis. **Third-degree burns** destroy skin tissues down to and including the subcutaneous layer.

    A nurse who works in the burn ward of a local hospital notices that patients who have second-degree burns suffer more pain than patients who have third-degree burns.

    Using your knowledge of the skin, **explain** why the more severe burn causes less pain.

The Cardiovascular System

Objectives

After completing this exercise, you should be able to:

- diagram the pattern of blood circulation around the body and label the blood in each location as oxygenated or deoxygenated
- differentiate between the pulmonary and systemic circuits
- identify and explain the function of each of the major structures of the heart
- discuss the connection between blood vessel diameter and the rate of blood flow
- discuss the effect of exercise on pulse rate, blood pressure, and the efficiency of oxygen transport
- discuss the ways in which exercise changes the distribution of blood in the body
- apply your knowledge of vessel and heart structure to cardiovascular health issues

CONTENT FOCUS

We hear a lot about exercising our muscles and keeping fit. Activities that increase cardiovascular fitness are an important part of any exercise program. Many people don't realize that the heart is a muscular pump, and so, as with all our body muscles, heart function can be improved by exercise and activity.

The average heart is only about the size of your fist. It normally beats about **72 times per minute.** This may not sound too impressive, but each day, the heart pumps **1500 gallons** of blood. Over your lifetime, this adds up to enough blood to fill **13 supertankers!**

Each beat of the heart **transports blood** that contains **food, oxygen,** and **hormones** to all the cells of the body and **removes wastes** such as **carbon dioxide.**

The cardiovascular system of humans and most other animals consists of **three basic elements: a pump, blood vessels,** and **blood (the circulatory fluid).** In this exercise, you'll take a closer look at the heart and do some experiments to demonstrate how the circulatory system adjusts to the needs of the body.

ACTIVITY 1

CIRCULATION OF BLOOD AROUND THE BODY

The heart is divided into **four** chambers. The two **upper** chambers, called **atria,** receive blood returning to the heart. The two **lower** chambers, called **ventricles,** pump blood out.

1. The circulatory pathway that carries blood through the lungs and back to the heart is called the **pulmonary circulation.** The circulatory pathway that carries blood through the upper and lower body and back to the heart is called the **systemic circulation.** Following these pathways, you can see that **blood returns to the heart TWICE as it circulates around the body.**

   As shown in **Figure 15-1, blood returns from the body** into the **right atrium.** The blood then travels into the **right ventricle,** which pumps it out to the **lungs.** Blood **returns from the lungs** into the **left atrium** and then goes into the **left ventricle,** which pumps it to the **rest of the body.**

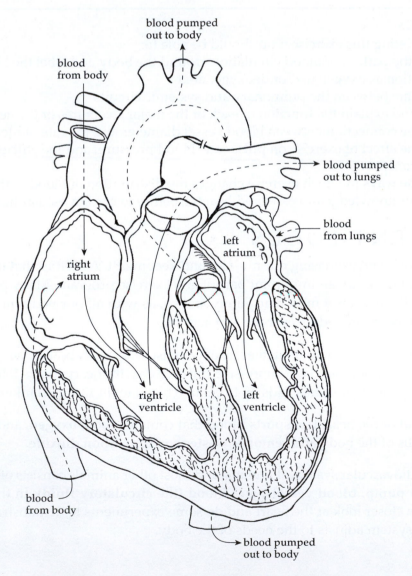

FIGURE 15-1.   Circulation of Blood Through the Heart

2. Using a **colored pencil or highlighter** (and the information you've studied about lung function), color each **heart chamber in Figure 15-1** that contains **deoxygenated** blood (blood **low** in oxygen) in **blue.** Color each **heart chamber** that contains **oxygenated** blood (blood **high** in oxygen) in **red.**

3. Within the pulmonary and systemic circulatory systems, the blood travels through **arteries, veins,** and **capillaries.**

    Arteries always transport blood **away from the heart. Veins** carry blood **toward the heart. Between the arteries and veins** lie **beds of capillaries** where cells pick up oxygen and release carbon dioxide.

    On **Figure 15-2, place a letter "A" next to** each blood vessel that is an **artery.** Place a **letter "V"** next to each **vein.** Place a **letter "C"** next to each **capillary bed.**

4. On **Figure 15-2,** color each **heart chamber and blood vessel** that contains **deoxygenated** blood in **blue.**

    Color each **heart chamber and blood vessel** that contains **oxygenated** blood in **red.**

5. Using the **clues** below, correctly **label** the listed blood vessels on **Figure 15-2.**

BLOOD VESSEL	CLUE FOR IDENTIFICATION
Superior vena cava	Vessel that returns blood from the head to the heart
Inferior vena cava	Vessel that returns blood from the lower body to the heart
Pulmonary arteries	Vessel that carries blood to the lungs
Pulmonary veins	Vessel that returns blood from the lungs to the heart
Aorta	Vessel that distributes oxygenated blood around the body
Carotid arteries	Branch off the aorta that delivers blood to the head
Jugular veins	Vessel that carries deoxygenated blood from the head to the superior vena cava

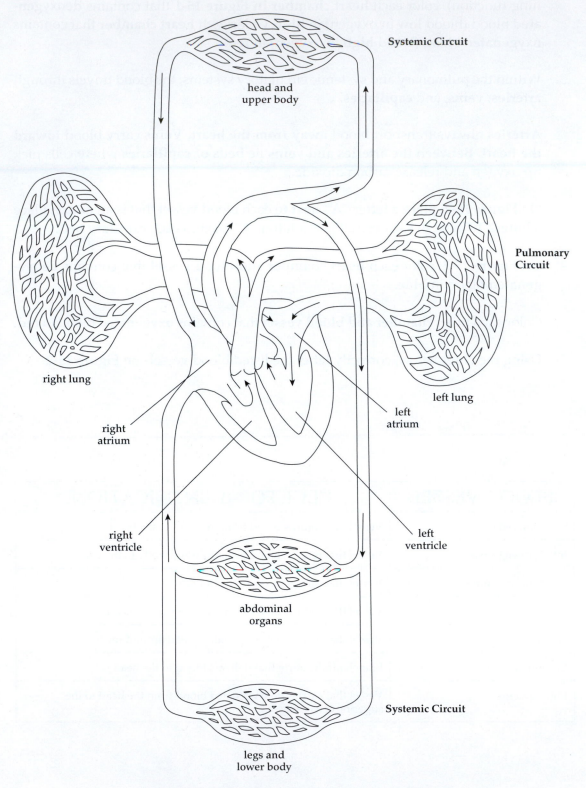

FIGURE 15-2. Pulmonary and Systemic Circulation

✓ Comprehension Check

1. **(Circle one answer.)** A drop of blood in the **pulmonary artery** is **oxygenated /
    deoxygenated.**

    Which heart chamber did it come from? _____

    Where will this drop of blood be found next? _____

2. **(Circle one answer.)** The blood flowing through the **left side** of the heart is
    oxygenated / deoxygenated. Explain your answer.

3. **(Circle one answer.)** A drop of blood in the **inferior vena cava** is **oxygenated /
    deoxygenated.**

    Which heart chamber will it enter next? _____

    Where will this drop of blood travel next? _____

4. Do all arteries carry **oxygenated blood? Explain** your answer.

**Check your answers and your Figure 15-2 labels with your instructor
before you continue.**

ACTIVITY 2   A CLOSER LOOK AT THE HEART

1.   Work in groups. Get a **model of the human heart.** Using **Figure 15-3** as a guide, locate the **four chambers** on the **heart model. Observe** the muscular walls of the left and right ventricles on the heart model.

     On which side of the heart is the muscular wall **thicker?** _____
     Why?

Hint:
Review the information in Figures 15-2 and 15-3.

---

2.   **Heart valves** prevent blood from flowing **backward** through the heart. Each valve section is anchored to the heart wall by **tendinous cords.** These strong cords hold the **valve sections closed** when the ventricles contract. When the cords are **relaxed,** the valve **opens.**

     On your model of the heart, locate **one valve between the right atrium and the right ventricle** (the **tricuspid valve**) and a **second valve** between the **left atrium and the left ventricle** (the **bicuspid** or **mitral valve**).

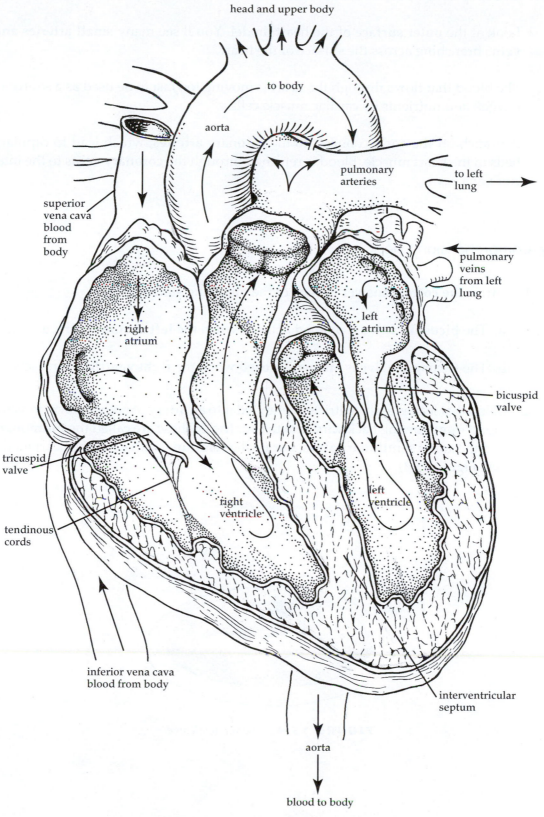

FIGURE 15-3. Interior View of the Heart

3.   To beat properly, the muscles of the heart must get an adequate blood flow.

     Look at the **outer surface** of the heart model. You'll see many **small arteries and veins** branching across the surface of the heart.

     The blood that flows through the heart is moving too fast to be used as a source of oxygen and nutrients by cardiac muscle cells.

     A branch off the **aorta** connects to the **coronary arteries,** which lead to capillary beds in the heart muscle. Blood is returned through the **coronary veins** to the inferior vena cava.

✔ Comprehension Check

1.   **Circle the correct answer** in each of the following statements.

    a. The **bicuspid** valve is **opened / closed** when the **left atrium** contracts.

    b. The **tricuspid** valve is **opened / closed** when the **right ventricle** contracts.

2.   Blood that leaves each ventricle is pumped into an artery. Another valve prevents blood from flowing backward into the ventricle. These are called the **pulmonary** and the **aortic semilunar valves. Locate** the **semilunar valves** on the heart model (see **Figure 15-4**).

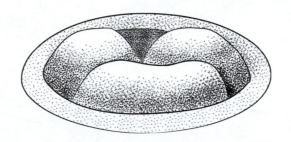

FIGURE 15-4.   Semilunar Valve

3. The semilunar valve on the **right** side of the heart is between **which heart chamber and which blood vessel?**

    _____   _____

4. The semilunar valve on the **left** side of the heart is between **which heart chamber and which blood vessel?**

    _____   _____

5. **(Circle one answer.)**

    a. When the right **ventricle** is **contracting,** the **pulmonary** semilunar valve is **opened / closed.**

    b. If a drop of blood is moving **toward the lungs,** it has passed through the **pulmonary / aortic** semilunar valve.

6. **Challenge Question!** When the right **atrium** is **contracting,** the **pulmonary** semilunar valve is **opened / closed.**

Check your answers with your instructor before you continue.

ACTIVITY 3   BLOOD VESSEL DIAMETER AND BLOOD FLOW

The ventricles contract to pump blood through all the arteries of the body. As blood travels away from the heart, it passes through smaller and smaller blood vessels.

How hard does the heart have to work to pump blood through blood vessels of different sizes?

How much force the heart exerts to pump blood **depends on how easily fluid passes through the blood vessel.** Using **rubber tubing to simulate blood vessels,** you'll design an experiment to determine whether the **diameter of blood vessels** affects the **rate of blood flow.**

During your experiment, you want to consider **only** the effect of blood vessel diameter. For this reason, you **won't** use a pump to simulate heart action. Blood will flow through your simulated blood vessels using only the force of **gravity.**

1. Work in groups.

    Get the following supplies: **a large container with spigot, a stopwatch, a plastic bucket, a large graduated cylinder, and three pieces of plastic tubing** (see sizes below):

    Large   6/16″ inside diameter

    Medium   5/16″ inside diameter

    Small   4/16″ inside diameter

2. **Develop a hypothesis** about the relationship between the **blood vessel diameter** and the **rate of blood flow.** Write it in the space below.

BLOOD FLOW HYPOTHESIS

 Check your hypothesis with your instructor before you continue.

3. Carefully **develop a plan** about how you'll get the information you need to test your hypothesis. **List the steps** of your plan, **including** the **equipment** you intend to use.

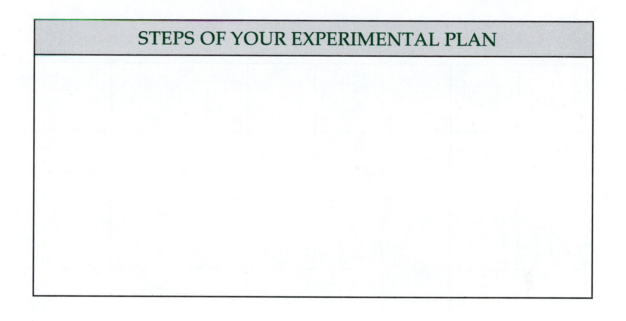

STEPS OF YOUR EXPERIMENTAL PLAN

4. In the space below, **make a table or chart** that shows your results clearly and neatly. Collect your experimental data and **record the results.**

5. **Plot a graph of your results in Figure 15-5.**

FIGURE 15-5. _____

6. Discuss the results with your group members. Write a couple of sentences that **summarize** your **results**.

Hint:

Remember—results include only facts, never opinions.

---

7. **Write a conclusion based on your hypothesis and collected data. Support your conclusion** by mentioning facts collected during your experiment.

8. **(Circle one answer.)** As blood travels from arteries into arterioles, the rate of blood flow will **increase / decrease.**

9. Capillaries are the **smallest** blood vessels. How does this fact affect the **rate** of blood flow?

10. Why is **flow rate** an important factor for determining how **efficiently** cells can pick up oxygen and remove carbon dioxide?

11. In the disease **atherosclerosis,** fatty plaques accumulate on the inside of artery walls. What effect would this have on the **rate** of blood flow? **Explain** your answer.

Check your answers with your instructor before you continue.

ACTIVITY 4   THE EFFECT OF EXERCISE

Every time the ventricles contract **(systole),** forcing blood out of the heart, pressure in the blood vessels **increases.** When the ventricles **relax (diastole),** pressure in the blood vessels **drops** again. The pressure exerted by the blood on the walls of the blood vessels is called **blood pressure.** In the original instrument used to measure blood pressure, a **column of mercury** was forced upward in a glass tube. Newer instruments that measure blood pressure still use the same measurements (**millimeters of mercury** or **mm Hg**). The average blood pressure in a young adult male is approximately **120 mm Hg during systole** and **80 mm Hg during diastole.** This is expressed as **120/80.** In young adult females, the average is about 8–10 mm Hg less than in males. Blood pressure can be lower in adults who exercise regularly.

Blood is forced into your arteries during each contraction of the heart. You can feel this wave of blood moving through the arteries as a **pulse** in the carotid artery of your neck. The pulse can also be felt in the radial artery of your wrist and other locations in the body. The **number of pulses** tells you **how fast the heart is beating.** The average pulse rate in a young adult is **72 beats per minute.**

1. Work in **groups.** Do the following experiments.

    One student will be the test subject. Others will monitor blood pressure, pulse rate, serve as timekeepers, and record the experimental data.

Caution!
DON'T be the test subject for this activity if you have heart or blood pressure problems!

2. Get the following supplies: **a blood pressure monitor, a stopwatch, and an aerobic step.**

3. Sit quietly for **one minute. While sitting,** take your **resting pulse rate and blood pressure.**

   Resting pulse rate: _____

   Resting blood pressure: _____

4. **Rapidly step up and down** on the aerobic step for **five minutes** (or do another form of exercise, as specified by your instructor). You should really be working out!

   Immediately sit down and record your **pulse rate and blood pressure.**

   Exercise pulse rate: _____

   Exercise blood pressure: _____

5. Is there a difference in pulse rate before and after exercise? _____

   If so, **describe the difference.**

6. Is there a difference in blood pressure before and after exercise? _____

   If so, **describe the difference.**

7. What is the heart doing when **blood pressure** increases?

   What is the heart doing when the **pulse rate** increases?

SELF TEST

1. Follow a drop of blood around the body. Begin with the tissue capillaries supplying your left big toe. **Place these locations in the correct sequence.**

    __1__ Tissue capillaries in left big toe

    _____ Pulmonary artery

    _____ Left atrium

    _____ Pulmonary vein

    _____ Right atrium

    _____ Left ventricle

    _____ Inferior vena cava

    _____ Aorta

    _____ Right ventricle

    _____ Arterioles (small arteries)

    _____ Venules (small veins)

    _____ Lungs

2. For the following body locations, enter **"D"** if the blood is **deoxygenated** and **"O"** if the blood is **oxygenated.**

    _____ Tissue capillaries entering big toe

    _____ Pulmonary artery

    _____ Left atrium

    _____ Pulmonary vein

    _____ Right atrium

    _____ Left ventricle

    _____ Inferior vena cava

    _____ Aorta

    _____ Right ventricle

    _____ Tissue capillaries leaving big toe

    _____ Arterioles

    _____ Venules

3. If there were **no tendinous cords** attached to the tricuspid valve, and the right ventricle is contracted, where would the blood go?

4. If a person's pulse rate is 96 pulses per minute, what is this person's **heart rate?**

5. You're a doctor listening to heart sounds through a stethoscope. While listening, you notice an unusual "hissing" sound in a patient's heart. On consulting the patient's medical records, you find that as a child he suffered from rheumatic fever, which causes scar tissue to form around the heart valves. What might be causing the hissing sound?

6. When a person **blushes,** blood flow to the skin increases. In order for this change in blood flow pattern to occur, how must the **diameter** of skin arteries change?

7. Place an "**X**" in front of the **one** measurement of oxygen concentration **that's most likely to be correct for all four locations.**

X	RIGHT VENTRICLE	PULMONARY ARTERY	PULMONARY VEIN	LEFT ATRIUM
	40	40	100	100
	100	40	40	100
	100	100	100	100
	40	40	40	40
	40	100	40	100

The graph in **Figure 15-6** shows the distribution of blood (by percentage) in the cardiovascular system when the body is at rest and during heavy exercise.

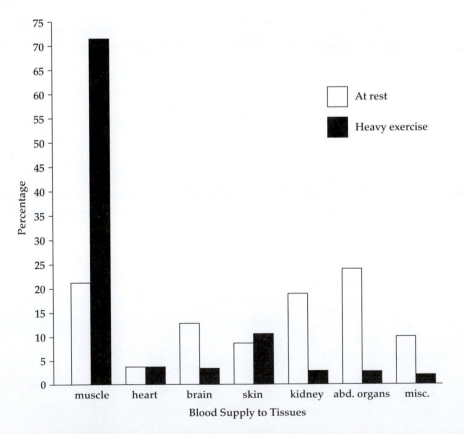

FIGURE 15-6.   Distribution of Blood at Rest and During Exercise

8. In reference to the graph in **Figure 15-6,** what **change** occurs to the **percentage of blood supply to the muscles** when you're exercising?

   Considering the **change in the percentage of blood flow** to the muscles shown in the graph, what probably happens to the **diameter of the blood vessels** that supply blood to the muscle tissue during exercise?

9. What **change** occurs in the percentage of blood supply to the **abdominal organs** during exercise?

   Considering the **change in the percentage of blood flow** to the abdominal organs shown in the graph, what probably happens to the **diameter of the blood vessels** that supply blood to the abdominal organs during exercise?

10. What **change** occurs in the **percentage of blood supply to the skin?**

    How is this **change** in blood flow **helpful to the body during exercise?**

Introduction to Anatomy: Dissecting the Fetal Pig

Objectives

After completing this exercise, you should be able to:

- identify and compare the external anatomical features of the male and female fetal pig
- identify and explain the function of each of the major structures of the mouth, the neck, and the thoracic cavity
- explain how organ structures are specialized to perform specific functions and give examples
- compare and contrast anatomical features of the pig with those of humans
- discuss the difference between lung capacity during normal breathing and deep breathing

CONTENT FOCUS

During the next several weeks, you'll be studying mammalian anatomy and physiology. To help you get a clear idea of the various organ systems and how they work, you'll be looking at the anatomy of the **fetal pig.**

Why fetal pigs? It may surprise you to learn that humans and pigs are very similar in anatomy. Today, pig organs and skin are frequently used for transplants and grafts to replace damaged human tissues. In addition, the skeleton of a fetal pig isn't fully calcified, making dissection easier to perform.

During extreme weather, farmers often sell surplus animals rather than run the risk of their dying from heat or cold. As part of the butchering process, all the organs, including the uterus, are removed. If fetal pigs are found in the uterus, they are preserved for educational purposes.

ACTIVITY 1   GETTING STARTED

1. Get the following supplies: **two long pieces of string, a tray, several paper towels, and some dissecting instruments (one scalpel, one pair of scissors, one large pair of forceps, and one blunt probe).**

2. Your laboratory instructor will distribute fetal pigs to each group or give you instructions for obtaining a pig.

3. Spread **two paper towels** on the tray. **Place the pig on its back on the tray.**

4. Tie the **ends** of the first string **tightly** around the two front hooves of your pig and slide it **underneath the tray.** Tie the **ends** of the other string around the two rear hooves and slide it underneath the tray.

5. Rotate the tray so that the pig's **tail is facing you.**

    Where is the **left side of the pig?** _____

> ### Note:
> Keep the position of the pig in mind as you proceed with the dissection. The instructions refer to the pig's right and left sides, NOT your right and left sides.

ACTIVITY 2   FOLLOWING ANATOMICAL DIRECTIONS

Before we begin the dissection, it's necessary to understand some special terms that relate to anatomical locations and directions in the pig's body. Using **Figure 16-1** as a guide, use the correct anatomical terms to describe the location of the following:

1. The pig, as placed on the tray, is lying on its _____ surface.

2. The umbilical cord is located on the _____ side of the pig.

3. The attachment of the umbilical cord to the belly is _____ to the **hind** legs.

4. The attachment of the umbilical cord to the belly is _____ to the **front** legs.

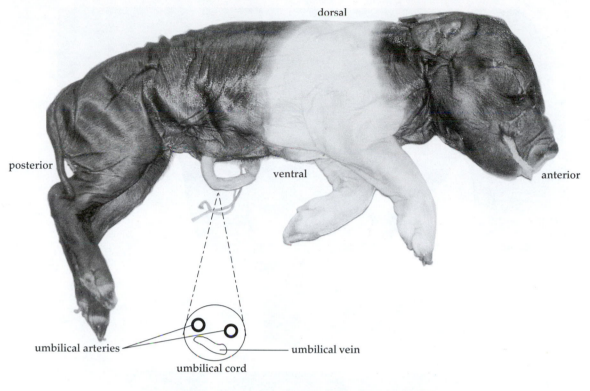

FIGURE 16-1. Anatomical Locations in the Pig

ACTIVITY 3   EXTERNAL ANATOMY OF THE FETAL PIG

1. Determine the **gender** of your pig, using **Figures 16-2 and 16-3** as a guide. There are two ways to tell males from females **externally.**

> ### Note:
> **Make sure you observe pigs of both genders.**

2. Look for the **anus,** which is located close to the base of the tail.

    Female pigs have a small, fingerlike projection, **just ventral to the anus.** This projection is called the **genital papilla.** On either side of the papilla, are the two **labia.** The **urogenital opening** is located between the two labia. The term **urogenital** refers to the double function of this opening. It allows for the excretion of urine (**"uro"** refers to the urinary system) and also leads to the **reproductive structures** (**"genital"** refers to the reproductive system).

> ### Hint:
> **If the papilla is not evident, you have a male pig.**

3. In **male pigs,** you'll notice the skin appears puffy or baggy just **anterior to the anus.** This is the **scrotum,** which contains the **testes.**

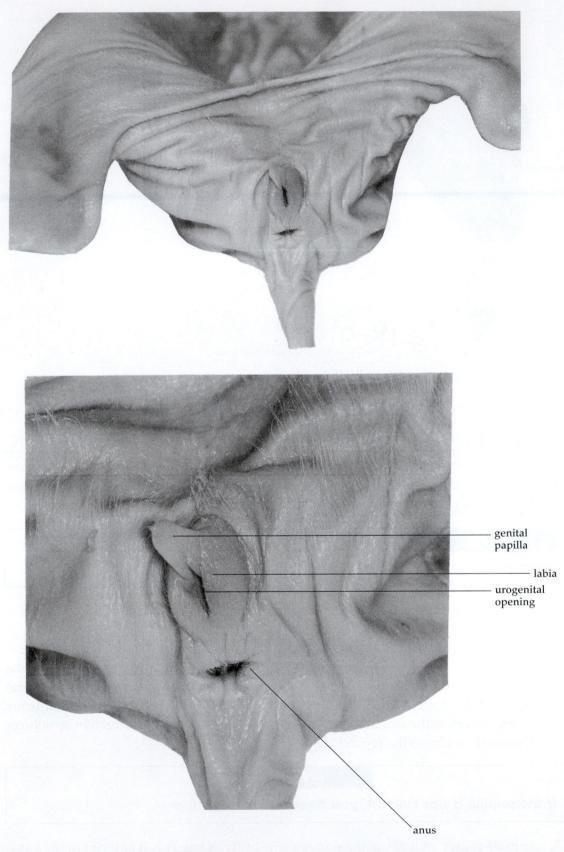

FIGURE 16-2. Female External Anatomy

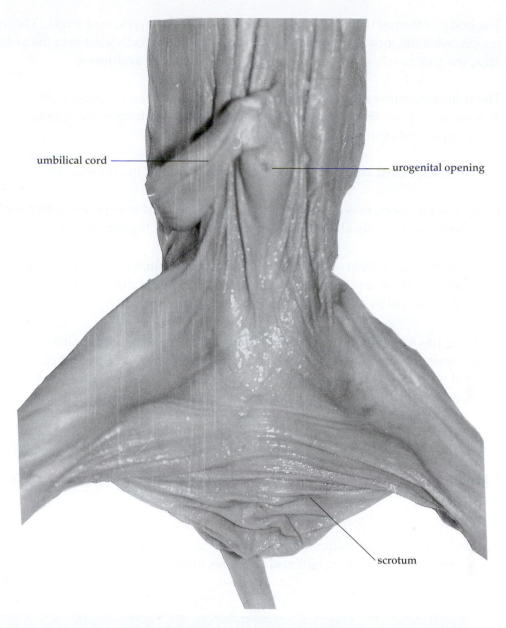

umbilical cord ——————————————— urogenital opening

scrotum

FIGURE 16-3. Male External Anatomy

4. The **urogenital opening** in the male is located just **posterior** to the attachment of the **umbilical cord** to the body wall. As with the female pigs, this opening functions in both the urinary and reproductive systems. Both urine and sperm are released here, but not at the same time.

    You can't see the **penis** on the exterior of the body because pigs, like many other four-legged animals, have a **retractable** penis, which can be seen when the animals reproduce. You'll notice that both male and female pigs have two rows of **nipples** on either side of the umbilical cord. As with humans, the nipples are functional only in females.

5. The body of the pig is divided into three regions: **head, neck,** and **trunk.** The trunk region, with the four legs and tail attached, is further subdivided into the **anterior** area, the **thorax** (chest area), and the **posterior** area, the **abdomen.**

    The thorax contains the heart and the lungs, enclosed by a protective rib cage. The abdomen contains the organs of the digestive system along with organs of many other important systems.

6. The **umbilical cord** attaches the fetus to the **placenta** in the mother.

    Using your scissors, make a cut across the umbilical cord to expose the two **umbilical arteries** and the **single, larger, umbilical vein** (see **Figure 16-1**).

    Push against the outside of the wall of the umbilical vein with a blunt probe. Repeat the process with an umbilical artery. Note that the wall of the umbilical vein is thinner and bends easily in comparison to the umbilical arteries.

7. Each pig has a **slit** cut in the side of the neck. This cut was made to inject the circulatory system with **colored latex** (a form of liquid rubber). The **arteries** have been injected with **pink latex** and the **veins** have been injected with **blue latex.** This will help you tell the blood vessels apart when you study the cardiovascular system.

ACTIVITY 4   DISSECTION OF THE MOUTH

1. Carefully cut through **both sides of the jaw,** as illustrated in **Figure 16-4.** As you cut, **alternate** between the right and left sides until you can open the mouth wide.

> ### Caution!
> Young pigs have very sharp, pointed teeth. Be careful where you put your fingers while opening the mouth.

2. Look at the **roof** of the mouth and **feel its texture.** The rippled **anterior region** is hard because it's composed of **bone** and **cartilage.**

    This area is called the **hard palate.** Notice that the hard palate is divided into **two halves.** The line separating the left and right sections of the hard palate runs directly along the **body midline.**

    Label the **hard palate** and the **body midline** on the diagram in **Figure 16-5.**

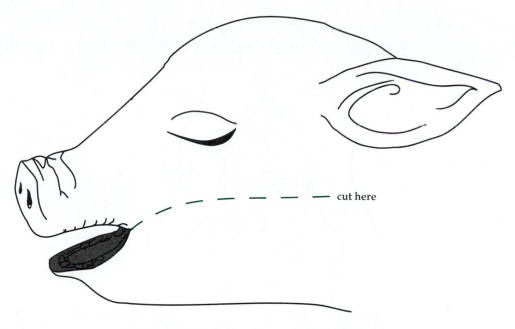

FIGURE 16-4.   Pattern for Jaw Incisions

3. The **posterior region** of the roof of the mouth is the **soft palate.** Feel the texture of this area. **How is it different** from the hard palate area?

4. The hard and the soft palates separate the mouth from the **nasal cavity.** As you may have experienced, drainage from the nose can enter the mouth, so there must be a **common area** where these two regions come together. This is the **pharynx.**

    The pharynx can be seen **posterior** to the soft palate.

    Insert the **tip of your blunt probe** into the opening of the pharynx and probe **anteriorly underneath the soft palate. Your probe is now in the nasal cavity.**

    Add labels for the **pharynx** and **soft palate** to Figure 16-5.

5. Pull the mouth completely open and look down into the throat. There's a hood-shaped flap of tissue called the **epiglottis.** The epiglottis prevents food and liquid from entering the **glottis,** the opening to the air passageways (the prefix **"epi"** means outside).

    Add labels for the **epiglottis** and **glottis** to **Figure 16-5.**

6. If you're eating and your food "goes down the wrong pipe," **which pipe** did the food enter? _____

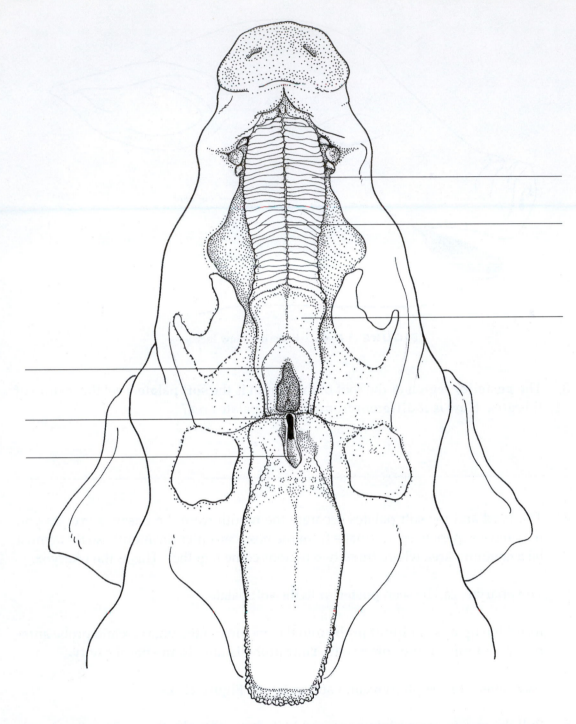

FIGURE 16-5. Structures of the Mouth

Check your Figure 16-5 labels with your instructor before you continue.

ACTIVITY 5   DISSECTION OF THE NECK

Caution!
While cutting, be careful to keep the point of the scissors away from underlying organs.

Referring to **Figure 16-6,** make **only** the following cuts:

1. Cut the skin and muscle layers between the chin whiskers **(number 1)** and the anterior part of the sternum or breastbone **(number 2)** as in **Figure 16-6.**

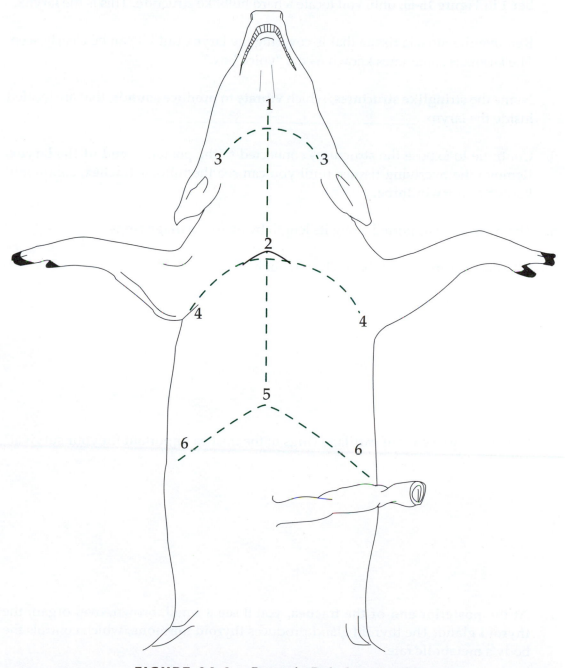

FIGURE 16-6.   Pattern for Body Cavity Incisions

2. Make two additional incisions **across the throat (number 1 to number 3)** as shown in the figure.

3. Use your **blunt probe** to separate the muscles from the organs in the neck region. You'll expose two long sections of the **thymus gland,** one on each side of the neck. Notice that the thymus is quite large (it covers most of the throat area). In a fetus, the thymus has an important role in **immune system function** and produces a hormone that stimulates immune system development. After puberty, the thymus begins to decrease in size and function.

4. With your fingers, feel around at the **anterior end** of the dissected area **(toward number 1 in Figure 16-6),** until you locate a hard bulblike structure. This is the **larynx.**

    Remove the muscle tissue that is covering the larynx until it can be clearly seen. The larynx is sometimes known as the **"voice box."**

    Name the **stringlike structures,** which **vibrate to produce sounds,** that are located inside the larynx. _____

5. Continue to expose the structures connected to the **posterior end of the larynx.** Remove the overlying tissues until you can see the tubular **trachea,** commonly known as the **windpipe.**

6. The trachea is supported along its length by many **cartilage rings.**

    What is the **function** of the trachea in your body?

    Why is the **presence of cartilage rings** in the trachea important for your survival?

7. At the **posterior end of the trachea,** you'll see a small, brown, oval organ, the **thyroid gland.** The thyroid gland produces **thyroid hormone,** which controls the body's **metabolic rate.**

8. Using your **blunt probe,** carefully separate the connective tissue from the sides of the trachea. The esophagus is **dorsal to the trachea.**

    Feel underneath the trachea and **lift the esophagus** with your probe. How is the esophagus **different from the trachea in structure?**

9. Is the esophagus part of the respiratory system? _____ **Explain** your answer.

Check your answers with your instructor before you continue.

10. On the side of the neck **opposite to the slit,** carefully remove the overlying muscle and connective tissue to expose the **carotid artery** and **two jugular veins.**

    These vessels will be running **parallel** to the trachea (see **Figure 16-7**).

11. The **carotid artery** is the **blood vessel closest to the trachea** and should be filled with **pink** latex. The **two jugular** veins should appear **blue.** The carotid artery carries blood rich in oxygen and nutrients toward the head. The jugular veins carry deoxygenated blood back toward the heart.

12. **Dissection Challenge!** Can you find the **vagus nerve** in the neck? It looks like a **flat,** white thread running close to the carotid artery and internal jugular vein. The vagus nerve is crucial for maintaining the normal state of homeostasis. Among other functions, this nerve **regulates heart rate, breathing, and digestive system activity.**

13. **Label** the following structures on **Figure 16-7: thymus, larynx, trachea, thyroid gland, carotid artery,** and **jugular veins.**

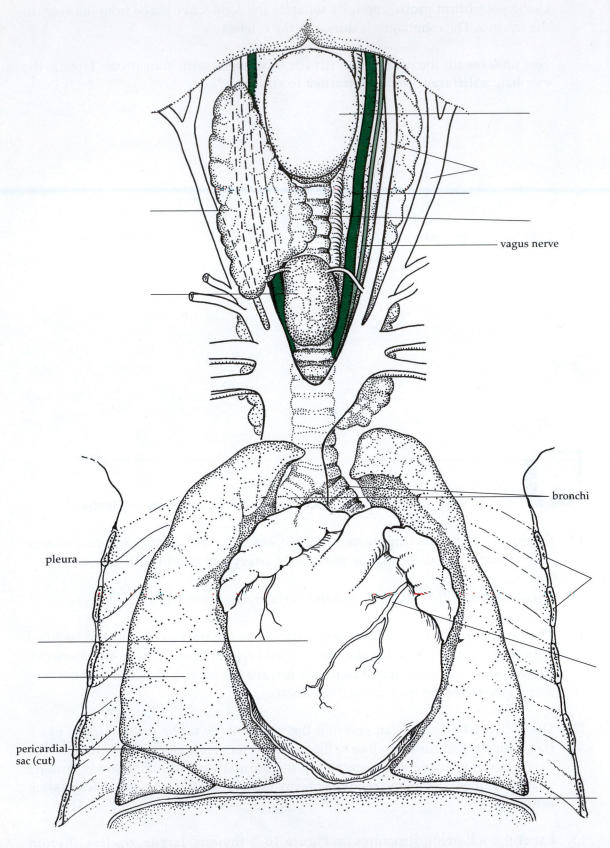

vagus nerve

bronchi

pleura

pericardial-
sac (cut)

FIGURE 16-7. Organs of the Neck and Thoracic Cavity

✔ Comprehension Check

Fill in the blanks with the choice that is **most appropriate** for the listed functions. **Answers can be used only once.**

a. carotid artery
b. esophagus
c. larynx
d. trachea
e. thyroid gland
f. vagus nerve
g. thymus
h. jugular vein

1. _____ Tube that transports food to the stomach.

2. _____ A blood vessel carrying deoxygenated blood back to the heart.

3. _____ This structure produces a hormone that controls the body's metabolic rate.

4. _____ Tube held open by cartilage rings.

5. _____ A blood vessel carrying nutrient-rich blood to the head.

6. _____ This structure contains the vocal cords.

7. _____ Plays an important role in immune system defense.

8. _____ Regulates heart rate, breathing, and digestive system activities.

ACTIVITY 6   DISSECTION OF THE THORACIC CAVITY

1. With your fingers, find the **anterior end of the sternum.** With your **scissors,** cut through the **sternum** and the **rib cage** until you reach the end of the rib cage (cut from **number 2 to number 5** and from **number 2 to number 4** on **Figure 16-6**). This will give you a clearer view of the ribs and help you pinpoint the site of your next incision.

2. **Make two more incisions** toward the sides of the body to completely open up the chest cavity **(number 5 to number 6).** As you cut, you'll be separating the **ribs** from the **diaphragm,** a large muscle that extends **completely across** the body cavity. **The diaphragm expands the chest cavity so that air flows into the lungs.**

3. The thoracic cavity is lined with layers of smooth membranes called the **pleura.**

    As you look into the thoracic cavity, the **two lungs** are easy to see. Each lung is enclosed in a **pleural cavity,** lined by the **pleural membranes.**

    The lungs are composed of millions of tiny sacs, which are the sites of gas exchange for the whole body. **Oxygen enters the lungs and carbon dioxide is removed.**

> ### Note:
>
> **The trachea extends posteriorly from the larynx and divides into branches called bronchi, which extend to the lungs. The branches can't be seen at this stage of the dissection, but we'll examine them later.**

4. Located between the two lungs is the **heart.** The heart is a muscular organ that **pumps blood to the lungs and the body tissues.** The heart of the pig is located in the center of the chest cavity. This is also true of the human heart (it is not on the left side as is commonly believed).

    The **pericardium** (also called the **pericardial sac**) is a tough membranous sheet that completely encloses the heart. **Remove the pericardial sac** so that you can get a better look at the heart.

    You'll see numerous pink lines running across the ventral surface of the heart. These are branches from the **coronary arteries** that supply oxygen-rich blood to the heart **(cardiac)** muscle. Blockage of blood flow through the coronary arteries or their branches can cause cardiac muscle cells to die from lack of oxygen (a **heart attack**). One way to reduce your risk of heart disease is choosing a diet low in saturated fat.

5. **Add the following labels** to the diagram in **Figure 16-7: lungs, heart, coronary artery, ribs,** and **diaphragm.**

    Check your Figure 16-7 labels with your instructor before you continue.

6. After you complete your dissection:

    ■ Get a **plastic storage bag** for your pig. Place a strip of tape near the bottom of the bag and write your group's names on the tape.

    ■ Slide the pig off the tray with the **string still attached** to the limbs.

    ■ Wrap the pig in **damp paper towels** and seal it in the storage bag. Your instructor will tell you where to store your pig.

    ■ **Wash and dry your trays and tools. Your instructor will give you specific instructions for disposing of any remaining pig tissues.**

ACTIVITY 7   MEASURING LUNG CAPACITY

Just looking at the lungs doesn't give you a very good idea of how lungs function. In the following activities, you'll experiment on yourself and your lab partners and develop some ideas about how respiratory function adjusts to your body's needs when you exercise.

The amount of air that enters and leaves your lungs each time you take a breath is called the **tidal volume.** In most men and women, the tidal volume is about **500 cc (half a liter).**

You can increase the amount of air inhaled and exhaled by deep breathing. The **maximum** amount of air that you can move in and out during a single breath is called the **vital capacity.**

1. Work **with a partner. Each** student will perform this experiment and record his/her **individual** results.

Caution!
DON'T use the balloon method if you have respiratory or heart problems!

2. Get **a metric ruler, a clamp, and one balloon for each person.**

Note:
Read through the directions COMPLETELY before beginning!

3. Measure your respiratory volume according to the following directions.

    a. Sit down and relax.

    b. **Inhale normally** and then **exhale only that normal breath** into the balloon.

Note:
If you have trouble inflating the balloon, stretch it several times and try again.

    c. **Immediately** twist the balloon several times and **clamp it shut** so that no air escapes.

    d. Place the tip of a pencil vertically onto the "zero" cm mark. Place the balloon on its **side** next to the pencil (**position A** in **Figure 16-8**).

Holding the balloon in position, move your pencil point from the zero mark and place it on the ruler at the right edge of the balloon (**position B** in **Figure 16-8**).

Record the diameter of the balloon in **Table 16-1.**

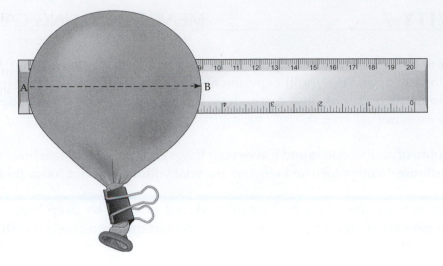

FIGURE 16-8. Method for Measuring Balloon Diameter

4. Repeat the entire process **twice more. Record** the results in **Table 16-1**.

T A B L E 16-1 MEASURING LUNG CAPACITY	
Diameter (cm) of Balloon Filled with Normal Breath:	
Trial 1	
Trial 2	
Trial 3	
Average	
Volume of air (cc) in balloon filled with normal breath:	
Diameter (cm) of Balloon Filled with Deep Breath:	
Trial 1	
Trial 2	
Trial 3	
Average	
Volume of air (cc) in balloon filled with deep breath:	

5. Calculate the **average balloon diameter** for your three trials of exhaled air **at rest.** Record the average in **Table 16-1.**

6. a. **Inhale as much air as you can** (with **only one** very deep breath) **and exhale as much air as you can** into the balloon. **Clamp and measure** the balloon as **in step #3** above.

   b. Record your results in **Table 16-1.**

7. Calculate the **average balloon diameter** for your three trials of exhaled air **while deep breathing. Record** the average in **Table 16-1.**

8. **Use the graph in Figure 16-9 to convert** the average diameter of the balloon filled by **a normal breath** and the balloon filled by **a deep breath** into **cubic centimeters (cc) of lung volume.**

   Record the information in **Table 16-1.**

9. What is your **tidal volume?** _____ cc

   What is your **vital capacity?** _____ cc

> ### Note:
> **Your measurement of tidal volume may be slightly higher than your actual tidal volume because you also exhaled the reserve air held in your lungs.**

---

10. Which volume of air is exchanged while you're sleeping? _____

    Which volume of air is exchanged while you're running? _____

    Explain your answer.

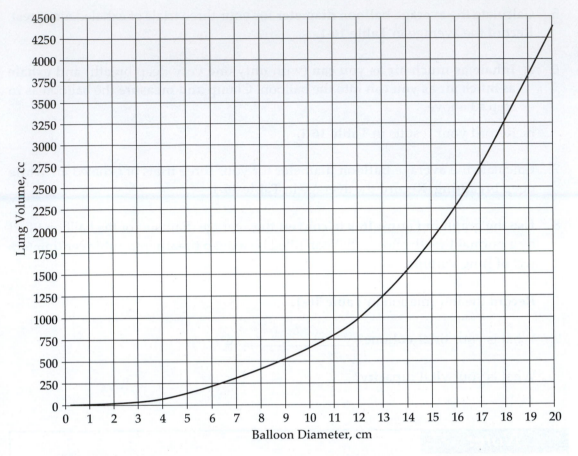

FIGURE 16-9.   Relationship Between Balloon Diameter and Volume of Exhaled Air

Check your answers with your instructor before you continue.

SELF TEST

Match the definitions in the right column with the terms in the left column. **Each answer can be used only once.**

1. _____ anterior   a. back

2. _____ dorsal   b. tail end of the body

3. _____ posterior   c. head end of the body

4. _____ ventral   d. belly side

5. Your lab partner arrived late to class on the day you started your fetal pig dissection. You've already selected a male pig for your dissection and you want your partner to choose a female. **Describe** how your partner can recognize and select a female pig from the container. **Be specific.**

6. Erin is playing shortstop on her college baseball team when she is hit in the throat by a line drive. On the way to the hospital, she has a very difficult time breathing. Which of the following is the most likely cause of her problem? **Explain** your answer.

   a. bruised diaphragm   d. crushed trachea
   b. concussion   e. blocked esophagus
   c. broken rib cage

7. Name **one part of the digestive system** that's located in the **thoracic** cavity.

8. Name a structure that lies **between** the thoracic and abdominal cavities.

9. When breathing, you're never able to completely fill your lungs with fresh air. A certain volume of stale air (with the oxygen removed) always remains in the air passageways. In some lung diseases, such as **emphysema,** large amounts of stale air accumulate and can't be expelled. The lungs are filled with air that is useless for gas exchange. **Vital capacity** is significantly less than in normal lungs.

    Predict what will happen when a person with emphysema exercises heavily. **Explain** your answer.

10. **Challenge Question!** In regard to your answer to question #9, explain how lack of air affects the body tissues. In your answer, use the following terms: **energy, oxygen, mitochondria, cell respiration, ATP,** and **red blood cells.**

EXERCISE

17

Organs of the Abdominal Cavity

Objectives

After completing this exercise, you should be able to:

- identify and explain the function of each of the major structures of the abdominal cavity
- explain how the stomach and small intestine are specialized to perform specific functions
- compare and contrast anatomical features of the pig with those of humans
- apply your knowledge of the benefits of increasing surface area in the radish root to similar modifications in the intestine and lung
- relate your observations of earthworm locomotion to peristaltic action in organs of the digestive tract

ACTIVITY 1   DISSECTION OF THE ABDOMINAL CAVITY

1. **Get your pig** from the storage container and set it up on a tray as you did in your previous pig dissection. Get a set of **dissecting instruments.**

2. Use your **scissors** to cut through the **skin and muscles** in the abdominal region.

   Pull up on the umbilical cord to hold the skin away from the abdominal organs and cut from **number 5 to number 7** (see **Figure 17-1**).

   Make two more incisions **around the umbilical cord (number 7 to number 8).**

293

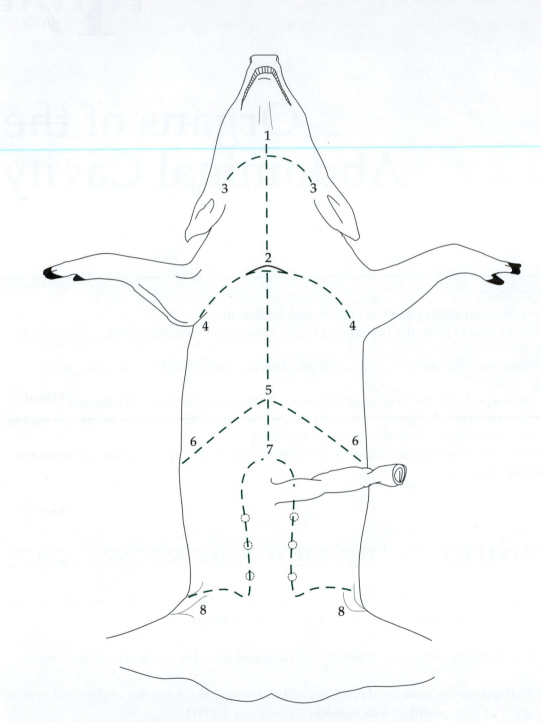

FIGURE 17-1.   Pattern for Body Cavity Incisions

3. The flap of tissue created by these cuts contains the **umbilical vein,** the **two umbilical arteries,** and the **urinary bladder.**

    Cut the umbilical vein **close to the umbilical cord.** This will allow you to turn back the flap so that it lies between the hind legs.

Hint:

If needed, rinse out the entire body cavity and blot it dry with paper towels. This will make it easier to locate and identify the abdominal organs.

---

4. Just as in the thoracic cavity, the abdominal cavity is sealed with a smooth membrane called the **peritoneum.**

5. The largest organ in the abdominal cavity is the **liver.** It's reddish-brown with several lobes and is located just **posterior to the diaphragm.**

    The liver has many functions including **storage of energy reserves, detoxification of poisons,** and **bile formation.**

6. And speaking of bile, lift the **right lobe** of the liver and locate the **gall bladder,** a small saclike organ.

    The gall bladder **stores bile produced by the liver. Bile aids in fat digestion.** Bile is released through the **bile duct** into the small intestine, where fat digestion takes place.

7. Raise the liver to locate the **stomach,** a large, hollow bag located on the left side of the body.

    The stomach stores food and releases it gradually into the small intestine. Digestion of proteins begins here.

8. The long, dark red organ lying to the left of the stomach is the **spleen.**

    The spleen **functions as part of the immune system** and also **removes damaged and worn-out blood cells from circulation.**

    Within the abdominal cavity, the organs are held in position by membranes called **mesenteries.** You can see this membrane attaching the spleen to the stomach.

9. The **pancreas** is located between the stomach and the small intestine. To expose this organ, lift the stomach and use your **blunt probe** to dissect through the membranes in this area. The pancreas is an **elongated, pale, granular** organ positioned **across the body cavity.**

    The pancreas **secretes digestive enzymes** through a duct into the small intestine. It also functions as an **endocrine organ,** producing **two hormones that control blood sugar level.**

10. The **small intestine,** on the right side of the abdominal cavity, is a long, thin tube. Most chemical digestion occurs here. In the small intestine, **enzymes break down proteins, fats, carbohydrates,** and **nucleic acids.**

    Hold up one of the loops of the small intestine and notice that it is held together by **mesenteries.** If you look closely, you can see **arteries** and **veins** fanning out across the membrane.

    Mesenteries provide a **support structure** for blood vessels and nerves leading to all the abdominal organs.

11. The **large intestine** is on the left side of the abdominal cavity. The main function of the large intestine is the **reabsorption of water from wastes,** preventing dehydration. The concentrated wastes are stored in the **rectum** and eliminated through the **anus.**

12. Lift the intestines on either side of the body and you'll see a **kidney.** The kidneys **remove nitrogen-containing wastes and form urine.**

13. The **urinary bladder,** located between the two umbilical arteries, is a **temporary storage organ for urine.** When the bladder fills, nerve endings in its muscular walls are stimulated, causing the muscles to contract for urination.

14. **Label** the following structures on **Figure 17-2: liver, gall bladder, stomach, spleen, pancreas, small intestine, large intestine, mesenteries, kidney,** and **rectum.**

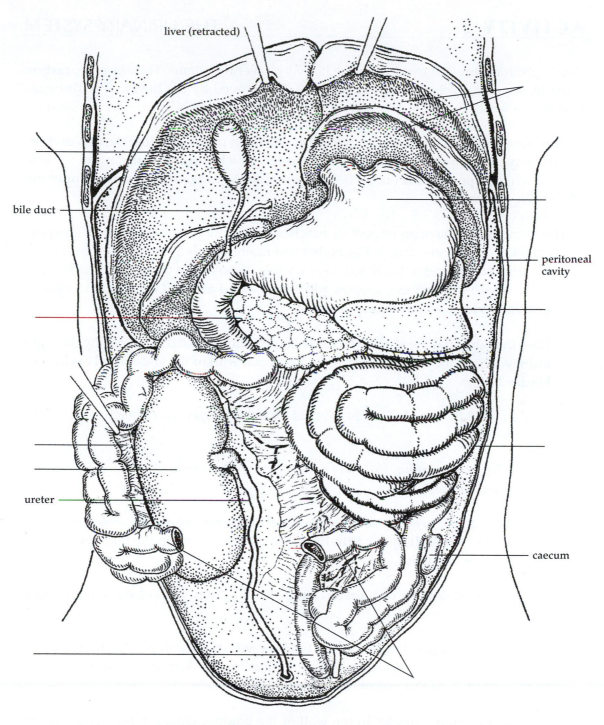

liver (retracted)

peritoneal
cavity

bile duct

ureter

caecum

FIGURE 17-2. Organs of the Abdominal Cavity

Check your figure labels with your instructor before you continue.

ACTIVITY 2   THE URINARY SYSTEM

Waste products are removed from the body by several systems. For example, **carbon dioxide,** a waste product of **cell respiration,** is removed by the **lungs.** The **nitrogen-containing** waste products of **protein digestion** are excreted by the urinary system.

The urinary system includes the two **kidneys,** which **produce urine;** the two **ureters,** which **carry the urine to the bladder;** the **bladder,** which **stores the urine;** and the **urethra,** which **transports urine** out of the body through the **urogenital or urinary opening** (depending on your sex).

The kidneys also have an important function in maintaining **homeostasis** (keeping internal body conditions stable). The body must maintain adequate levels of salts, sugars, and other substances in the bloodstream, while disposing of the excess. The kidneys play a key role in this process, keeping **pH** values, **fluid and salt content,** and **plasma levels of valuable materials** nearly constant.

1. Push the intestinal mass to the **left side** of the abdominal cavity, exposing the **right kidney.** The kidneys are located along the **dorsal wall of the abdominal cavity** (see **Figure 17-2**).

2. You'll notice that the kidney is covered by a thick layer of connective tissue, the **peritoneum.**

   Using the blunt probe, carefully remove the peritoneum, uncovering the right kidney and ureter.

   In addition to the ureter, you'll see the **renal artery and vein.** The **renal artery** carries blood **to the kidney for filtration.** The blood **returns** to the body circulation through the **renal vein.**

3. Each kidney is drained by a **ureter,** a tube leading from the kidney to the urinary bladder.

   Locate the ureter and follow it to the **bladder,** which is located in the flap of tissue created by your incisions around the umbilical cord in Activity 1. The bladder can be found **between the two umbilical arteries.**

   During urination, muscles in the wall of the bladder contract, forcing urine out of the bladder and into the **urethra,** a tube that transports urine out of the body. Urine is held in the bladder by **sphincter muscles** located where the bladder and urethra are connected.

   When the sphincter is open, urine passes from the bladder into the urethra. Certain health conditions affect the ability of the sphincter to prevent escape of urine **(incontinence).** Although incontinence can occur in adults of both genders, the condition occurs twice as often in women as in men. This is due in part to the structure of the female urinary tract, but pregnancy and childbirth also play a role.

4. Label the following structures on **Figure 17-3: kidney, urethra,** and **bladder.**

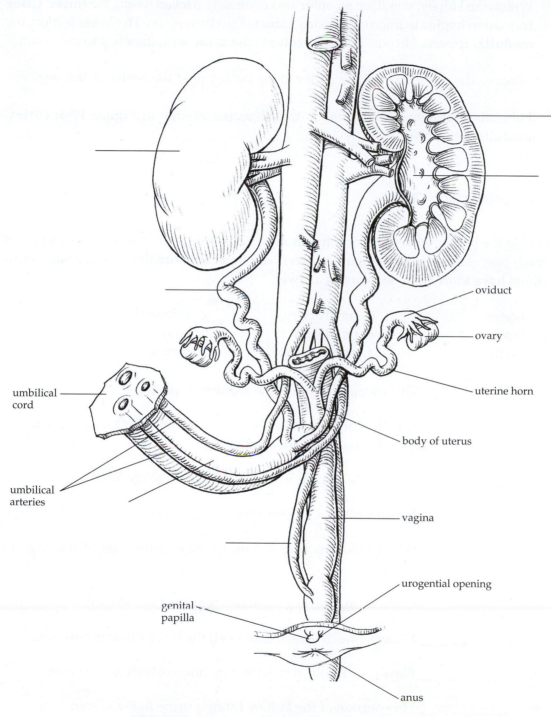

umbilical
cord

umbilical
arteries

genital
papilla

oviduct

ovary

uterine horn

body of uterus

vagina

urogential opening

anus

FIGURE 17-3.   Urinary System of the Female Pig

5. Cut the **ureter** and **renal blood vessels.** Remove the **intact right kidney** from the
    abdominal cavity. If necessary, remove additional peritoneum so that the kidney
    can be **removed without damage.**

6. Make an incision through the long axis of the kidney, cutting it into two equal halves (like opening a book).

Within the kidney, you'll see an **outer area** of densely packed tissue, the **cortex. Urine formation begins** in microscopic structures within this region. The **inner** section, the **medulla,** appears fibrous. This area **collects the urine** and funnels it to the ureter.

Observe the connection of the ureter to the kidney near the center of this region.

7. Label the following structures on the dissected kidney in **Figure 17-3: cortex, medulla,** and **ureter.**

Comprehension Check

Fill in the blanks with the choice that is most appropriate to describe the function of each part of the urinary system. **Answers can be used more than once. Some questions have more than one correct answer.**

a. ureter	d. renal artery	g. urinary bladder
b. kidney	e. renal vein	h. medulla
c. urethra	f. peritoneum	i. cortex

1. _____ This membrane holds the kidneys in position.

2. _____ If a drop of urine is in the ureter, which **TWO** structures will it pass through next **(in sequence)?**

3. _____ Carries blood high in metabolic wastes to the kidney.

4. _____ Returns filtered blood to body circulation.

5. _____ When this structure is full, nerve endings signal the urge to urinate.

6. _____ Urine formation begins in this region of the kidney.

7. _____ Either urine or semen could exit the body through this tube.

8. _____ Plays a vital role in maintaining homeostasis in the body.

9. _____ This portion of the kidney funnels urine to the ureter.

Check your answers and Figure 17-3 labels with your instructor before you continue.

ACTIVITY 3   INTERNAL STRUCTURE OF THE STOMACH

1. With your fingers, locate the **pyloric sphincter** of the fetal pig. It'll feel like a **small, hard mass** in the stomach wall. Using the same method, locate the **cardiac (gastro-esophageal) sphincter.**

    Referring to **Figure 17-4,** use your **scissors** to remove the entire stomach, being careful to include **both sphincters.**

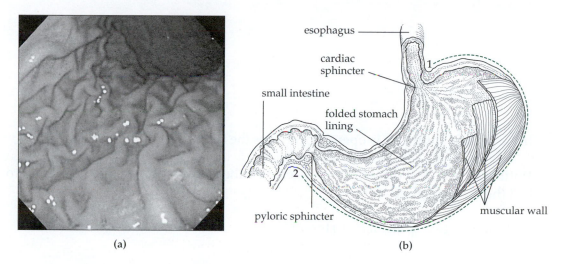

(a)   (b)

FIGURE 17-4.   (a) Adult Stomach Folds and (b) Internal Structure of the Stomach

2. Following the **dotted line** in **Figure 17-4,** cut from **number 1 to number 2** and open up the stomach.

    Rinse the inside of the stomach and **blot it dry** with paper towels.

3. Examine the **interior** of the stomach. **Observe** that the stomach wall is folded, like an accordion.

    Imagine that, disregarding what you know about dietary guidelines, you stuff yourself with two cheeseburgers, a supersize order of fries, an apple pie, and a chocolate shake.

    During this intake, what happens to the **size** of your stomach? _____

    How are the **folds in the stomach wall** related to this change in size?

The stomach has **two sphincters.** One is between the esophagus and the stomach (the **cardiac sphincter**) and the other is between the stomach and the small intestine (the **pyloric sphincter**).

Circle one answer: If there's food in the esophagus, the cardiac sphincter will be **opened / closed.**

What happens to the food in the **stomach** while **both sphincters are closed?**

If the **pyloric sphincter is open,** where does the food go? _____

4.   With one hand, feel the pyloric sphincter. At the **same time,** use the other hand to feel the cardiac sphincter.

How are the two sphincters **different?**

Which sphincter do you think is **stronger?** _____

The **cardiac sphincter** prevents food in the stomach from backing up into the esophagus.

The **pyloric sphincter** controls the **rate** at which food enters the small intestine. No matter how full your stomach is, food is always released **gradually** into the small intestine.

Comprehension Check

1.   Considering your examination of the stomach, **explain** what happens during **vomiting.**

2.   What would happen if your **pyloric sphincter** weren't able to **close?**

3.   What would happen if your **pyloric sphincter** weren't able to **open?**

4.   What would happen if the **cardiac sphincter** didn't close completely and stomach juices were allowed to enter the esophagus?

5.   **Challenge Question!** What other organ in the abdominal cavity has folds similar to those in the stomach wall? Why are these folds necessary for its function?

Check your answers with your instructor before you continue.

ACTIVITY 4   EXAMINING THE SMALL INTESTINE

1. Hold up a loop of the small intestine and **examine the mesenteries.**

    In addition to holding the small intestine in position, the mesenteries provide a passageway for a **large number of blood vessels.**

2. Using your **scissors,** carefully **snip through the mesenteries** and start separating the loops of the small intestine.

    Remove the small intestine.

3. Lay the intestine out on the laboratory counter and use a **meter stick** or **yardstick** to measure the length.

    The length of my pig's small intestine is _____.

    Circle one answer: The small intestine is **shorter / longer** than I expected.

    If the small intestine is longer than you expected, you might be interested in the origin of the name "small" intestine. The intestine is named for its **small diameter (and not its length).**

    The large intestine has twice the diameter of the small intestine **(but only half the length).**

4. Although each person is slightly different, the small intestine in humans averages about **20 feet (6 meters)** in length.

    Using your knowledge of the digestive system, **explain** how increasing the length of the small intestine would help this organ **perform its functions.**

5. Remove approximately **two inches (5 cm)** from the small intestine. Using your **scissors,** cut the piece of intestine **open lengthwise** and place it on the stage of the **dissecting microscope.**

6. Observe that the inside of the small intestine is **not smooth.**

    The interior wall is **folded,** and **each fold is carpeted** with slender fingerlike projections called **villi** (singular is **villus**).

7. After you've completed your dissection, do the following:

    ■ **Remove the strings** from the limbs.
    ■ Your instructor will tell you how to dispose of your pig and any remaining tissues.
    ■ Wash and dry all instruments and trays.

ACTIVITY 5   THE IMPORTANCE OF SURFACE AREA

1. The need to increase surface area is common throughout the plant and animal kingdoms. On the laboratory counter, you'll find a demonstration of a plant root **(in a radish seedling).** In this activity, you'll compare structures that increase surface area in a plant root with comparable structures that increase surface area in the small intestine.

2. The many slender extensions you can see covering the root of the radish seedling are called **root hairs.** Would a root without root hairs be able to absorb as many molecules from the soil as one with them? **Explain** your answer.

3. The interior wall of the small intestine is **folded** (as you observed with the dissecting microscope), and **each fold is carpeted** with slender fingerlike projections called **villi** (singular is **villus**). Although you can't see them at lower magnifications, each of the villi is covered, in turn, with even more tiny projections called **microvilli** (see **Figure 17-5**). Located inside each of the villi is a network of capillaries and lymphatic vessels **(lacteals).** These vessels pick up nutrients as they are absorbed, so that they can be distributed to all parts of the body.

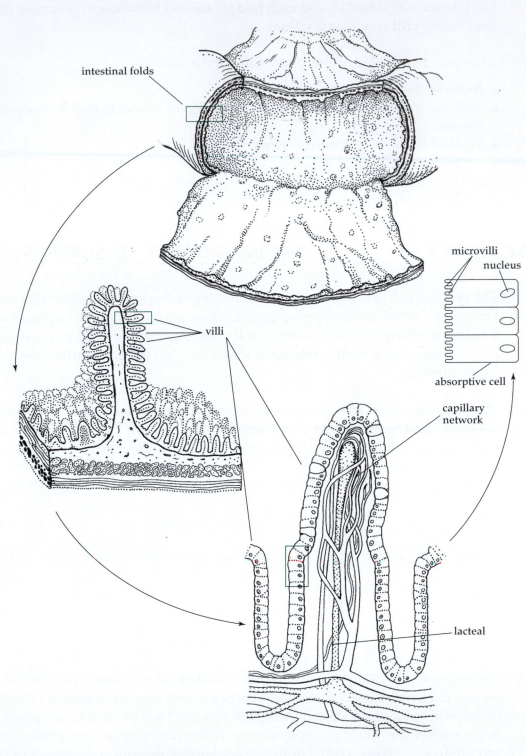

intestinal folds

villi

microvilli

nucleus

absorptive cell

capillary
network

lacteal

FIGURE 17-5.   Interior of the Small Intestine

The small intestine is quite a narrow tube. However, if the villi and the microvilli in the intestinal tract were laid out flat, the surface area would be **half the size of a basketball court!**

Why does the small intestine need such a large surface area?

✔ Comprehension Check

1. Which type of intestinal lining (with or without villi) can absorb the **greater number of nutrient molecules at one time? Explain** your answer.

2. Considering your answer to the previous question, explain how **microvilli** are beneficial for small intestine function.

Check your answers with your instructor before you continue.

ACTIVITY 6
A CLOSER LOOK AT THE INSIDE OF THE SMALL INTESTINE

1. Referring to **Figure 17-6,** locate the **villi** (projecting toward the center of the tube).

2. At the base of the villi, you'll see many **small,** circular structures.

    These are the **intestinal glands.** Intestinal glands produce **digestive enzymes** that perform the final breakdown of food molecules.

    Add a **label** for **intestinal glands** to the diagram in **Figure 17-6.**

3. Opposite the intestinal glands, you'll see a cluster of **larger** circles.

    These are **lymphoid nodules,** small lymph nodes that act as **bacterial filters** in the digestive, respiratory, and urinary tracts.

    Which other body locations contain **lymph nodes?**

    Add a **label** for **lymphoid nodules** to the diagram in **Figure 17-6.**

4. You can see **two layers of muscle tissue** around the outside of the intestine.

    The **innermost** layer of muscle fibers runs in a circle around the tube and is referred to as **circular muscle.**

    The **outermost** layer has the muscle fibers arranged along the **length** of the tube and is called **longitudinal muscle.**

    Add **labels** for the **circular and longitudinal muscle** layers to the diagram in **Figure 17-6.**

5. The **circular and longitudinal** muscle layers **alternately** contract, producing mixing movements and moving food through the small intestine with waves of **peristalsis.**

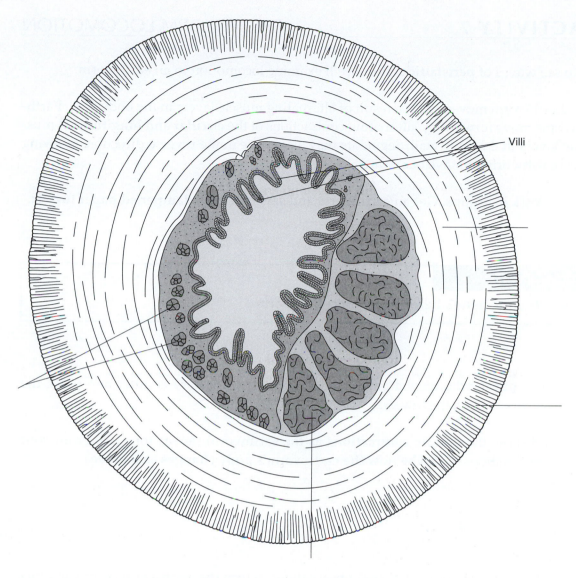

Villi

FIGURE 17-6. Cross Section of the Small Intestine

Check your figure labels with your instructor before you continue.

ACTIVITY 7   EARTHWORM LOCOMOTION

To see waves of **peristaltic action,** we'll examine locomotion in an earthworm.

Earthworm movement is a result of alternating muscle layers in the body wall. Earthworms move forward by alternating contractions of the **circular and longitudinal muscle** layers in the body wall. Segments may also move in the reverse direction, allowing the earthworm to move backward.

1.   Work in groups. **Get an earthworm and several clean, dampened paper towels.**

> ### Note:
> **Wash your hands before touching the worms!**

2.   **Lay out** the dampened paper towels on your laboratory counter and **place the earthworm** on the towels.

Observe the **muscle contractions** as the earthworm moves forward. From your observations, **describe how the earthworm's body changes** as it moves.

When you've completed your observations, **return the earthworm** to the container on the supply table.

3.   Considering your earthworm observation and the diagram in **Figure 17-7,** explain how the **process of peristalsis is** related to the earthworm's locomotion.

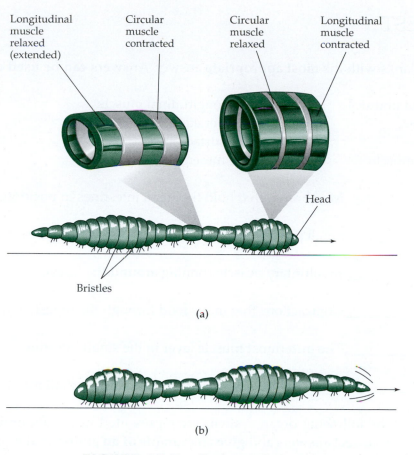

Longitudinal muscle relaxed (extended)

Circular muscle contracted

Circular muscle relaxed

Longitudinal muscle contracted

Head

Bristles

(a)

(b)

FIGURE 17-7. Earthworm Locomotion

4. **(Circle one answer.)** The circular and longitudinal muscles in the small intestine are composed of **smooth muscle tissue / skeletal muscle tissue / cardiac muscle tissue.**

SELF TEST

Fill in the blanks with the **most appropriate** answer. **Answers can be used only once.**

a. lymphoid nodules   e. longitudinal muscle
b. root hairs   f. peristalsis
c. villi   g. surface area
d. circular muscle   h. mesenteries

1. _____ Membranes that hold the small intestines in position.

2. _____ Small projections that increase surface area in the small intestine.

3. _____ Involuntary muscle running **around** the intestine.

4. _____ Contractions that move food through the digestive tract.

5. _____ The **outermost** muscle layer in the small intestine.

6. _____ Has a **function similar** to the villi of the small intestine.

7. Which of the following organ systems are represented in the **abdominal cavity?** Circle ALL correct answers and **give an example of an abdominal organ or structure** that belongs to each system.

ORGAN SYSTEM	EXAMPLES
a. Digestive system	
b. Respiratory system	
c. Urinary system	
d. Circulatory system	

8. When you experience **heartburn,** which **stomach** structure isn't functioning correctly? **Explain** your answer.

9. **Bulimia** is an eating disorder that involves forced vomiting. Over a period of years, bulimics gradually lose the enamel coating from their teeth. Using your knowledge of the digestive system, **explain** what causes the loss of tooth enamel.

10. **Urinary incontinence** (inability to control urine flow) frequently occurs in the elderly. Which organ of the urinary system may be malfunctioning? Which structure in the organ you named is malfunctioning? **Explain** your answer.

11. **Figure 17-8** contains two designs for air sacs in a human lung. Which will be **more efficient** in taking in oxygen and removing carbon dioxide? **Explain** your answer.

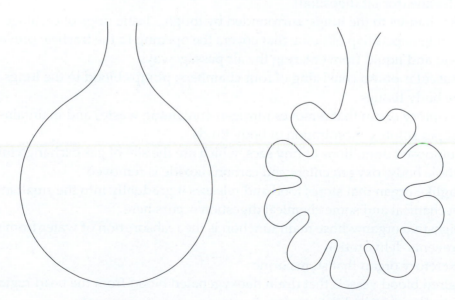

Air Sac Design A   Air Sac Design B

FIGURE 17-8.   Comparison of Air Sac Designs

Complete the crossword puzzle in **Figure 17-9** to review the structures of the pig and their functions.

ACROSS
2. Pair of blood vessels that carry oxygen-poor blood that is high in wastes from the fetus to the placenta; called the umbilical _____
3. Most chemical digestion occurs here; enzymes break down proteins, fats, carbohydrates, and nucleic acids
8. External genital structure in female pigs
11. Part of the immune system; site of T-cell development
13. Muscular tube that connects the mouth to the stomach
16. Part of the immune system; blood-filtering organ and storage site for red blood cells and some white blood cells
17. Thin connective tissue membranes that hold the internal organs in place and provide a passageway for blood vessels and nerves that supply the organs
18. Bony roof of the mouth
19. Houses the vocal cords, the vibrations of which make speech possible; composed of cartilage
20. Fleshy sac that contains the testes in males
21. Endocrine (hormone-producing) gland; produces a hormone that helps in regulating the metabolic activity of the body
22. Secretes digestive enzymes into the small intestine; also functions as an endocrine organ, producing two hormones that control blood sugar levels
23. Has many functions, including storage of energy reserves, detoxification of poisons, and bile formation

DOWN
1. Stores bile formed by the liver; releases it through the bile duct into the small intestine (for fat digestion)
4. Air channel to the lungs; surrounded by tough, elastic rings of cartilage
5. Hood-shaped flap of tissue that covers the opening to the trachea; prevents food and liquid from entering the air passageways
6. Muscular pump consisting of four chambers; pumps blood to the lungs and the body tissues
7. Excretory organ that removes nitrogen-containing wastes and maintains the proper solute concentration of body fluids
9. Composed of millions of tiny sacs, which are the site of gas exchange for the whole body; oxygen enters and carbon dioxide is removed
10. Baglike organ that stores food and releases it gradually into the small intestine; mechanical and some chemical digestion occurs here
12. Digestive organ whose main function is the reabsorption of water from wastes; prevents dehydration
14. Excretory organ that stores urine
15. Paired blood vessels that drain deoxygenated blood from the head region back to the heart; called the jugular _____

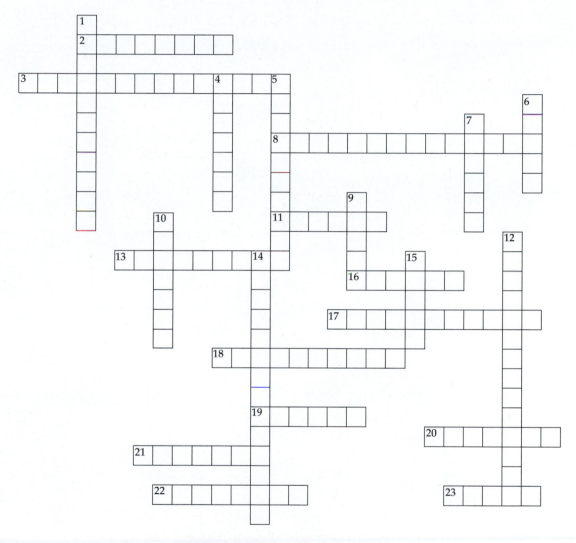

FIGURE 17-9. Crossword Puzzle

Introduction to Forensic Biology

Objectives

After completing this exercise, you should be able to:

- identify and interpret basic fingerprint patterns
- use your knowledge of fingerprint patterns and details to analyze forensic evidence
- demonstrate an understanding of the A–B–O and Rh blood typing systems
- use antigen–antibody interactions to identify and type unknown blood samples
- apply your knowledge of basic forensic techniques to real-life situations

CONTENT FOCUS

A scientist is similar to a detective. Both use scientific thinking to solve problems that have no clear solutions. Solving this type of open-ended problem requires you to develop your skills in forming hypotheses, gathering evidence by observation and experimentation, and drawing conclusions on the basis of your findings.

Forensic science is a good example of the practical application of the scientific method. In this case, science is used to analyze evidence and solve crimes. Forensic scientists use information from all branches of science, including chemistry, geology, physics, medicine, and, of course, **biology.**

This exercise provides an introduction to two of the basic techniques that are employed in forensic biology: **fingerprint analysis** and **blood typing.**

ACTIVITY 1   GETTING STARTED

Fingerprinting is the most commonly used and reliable system of identification. Each person has a unique pattern of ridges and valleys on his/her fingertips. Although other physical characteristics (such as weight, height, and hair color) may change as a person ages, fingerprints remain the same for life. Fingerprints penetrate through five skin layers and so are almost impossible to erase. Criminals have tried filing, acid burning, and even surgical removal, but with limited success.

Fingerprints have been used for identification purposes for several thousand years. The Babylonians recorded their fingerprints in the soft clay of their writing tablets (paper wasn't yet in use). The fingerprints served as a "signature" to prevent document forgeries. A similar system was also used in ancient China and Japan. In the past 100 years, an internationally recognized system of identifying fingerprints has been developed. Most law enforcement agencies (even in local communities) maintain their own fingerprint files. They can also access fingerprint information from state agencies and the FBI.

The skin is covered with **sweat glands,** including some in the ridges on the fingers and palms. During normal activities, perspiration from the sweat glands accumulates in these ridges. These ridges also accumulate body oils (from touching oil-producing areas such as the face, scalp, or neck). When you touch an object, perspiration and skin oils are transferred to that surface, leaving an invisible impression (the **latent fingerprint**). Latent prints found on nonabsorbent surfaces (such as metal or glass) can be dusted with colored powder and removed with transparent tape. Latent prints on absorbent surfaces (such as cloth or paper) must be recovered chemically.

The **Fingerprint Classification System** has three basic patterns: **arches, loops,** and **whorls.**

1. Begin by learning to recognize the basic identifying patterns used in fingerprint analysis.

Pattern area	Part of the fingerprint that is used for identification **(Figure 18-1a)**
Ridgelines	Individual lines that make up a fingerprint
Delta	Area where **two ridgelines** come to a **point (Figure 18-1b)**
Core	Approximate center of the finger impression **(Figure 18-1b)**
Whorl	**Complete circle** formed by ridgelines between two deltas; whorls **don't** continue off either side of the print **(Figure 18-2)**

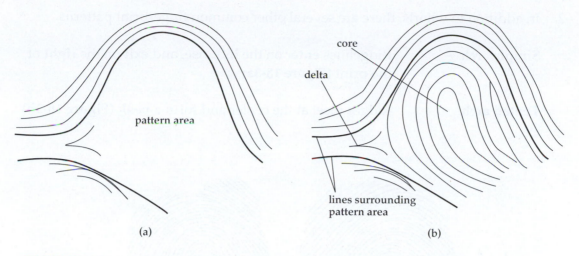

FIGURE 18-1. (a) Fingerprint Pattern Area and (b) Features Used in Identifying Fingerprints

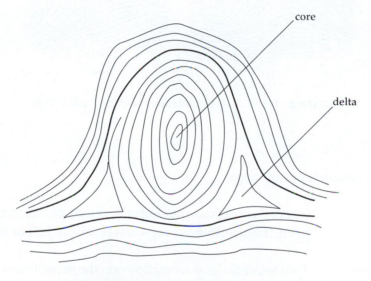

FIGURE 18-2. Fingerprint with Whorl

✔ Comprehension Check

1. Refer to **Figure 18-1b.** How many **ridgelines** are there between the delta and the core? _____

2. **(Circle one answer.)** The fingerprint in **Figure 18-1b is / is not** a whorl.

3. Refer to **Figure 18-2.** How many **ridgelines** are there between the delta and the core? _____

Check your answers with your instructor before you continue.

2. In addition to whorls, there are several other common fingerprint patterns.

    Simple arch   Ridgelines **enter** on the **left,** rise, and **exit** on the **right** of
                                 the print (**Figure 18-3a**)

    Tented arch   Ridges meet at the center and form a **peak** (**Figure 18-3b**)

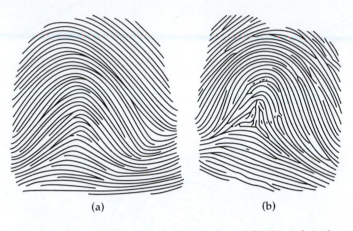

(a)   (b)

FIGURE 18-3.   (a) Simple Arch and (b) Tented Arch

3. There are various types of loops.

    Loops   Curved ridgelines that **enter** and **exit** on the **same side** of the
                                 print (**Figure 18-4a**); loops can be either **left or right** facing

    Double loop   Two separate loop formations on the same finger (**Figure 18-4b**)

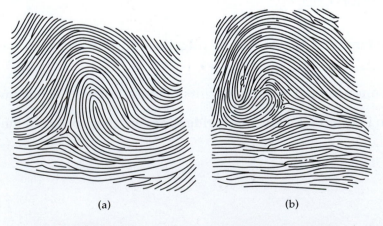

(a)   (b)

FIGURE 18-4.   (a) Loop and (b) Double Loop

4. The fingerprint classification system doesn't permit designating a fingerprint as more than one type. **For example, you can't designate the same fingerprint as both a loop and an arch.** Look at the fingerprint and identify the **largest, most dominant pattern.** This is the pattern you'll use to designate the print.

   Because you can't apply more than one fingerprint pattern to a single print, fingerprints that don't fit easily into any of the three categories are sometimes **grouped into a fourth category,** called **mixed.** Mixed prints usually show a combination of two or more different patterns.

5. Because there are only a few possible patterns, additional identifying characteristics are required to make a fingerprint match. A few examples are listed below:

Bifurcation	Forking or dividing of one ridgeline into two or more branches **(Figure 18-5)**
Divergence	Spreading apart of two ridgelines that have been running parallel or nearly parallel **(Figure 18-5)**

   Because scars and other marks that accumulate on your fingers as a result of daily wear and tear are unique, they are convenient features to use for identification. **Scars appear as white lines** on the fingerprint.

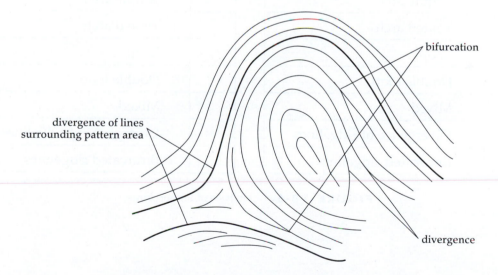

FIGURE 18-5.   Additional Identifying Features

✓ Comprehension Check

1. Place an **"X"** in front of each feature that is found in the fingerprints in **Figure 18-6.**

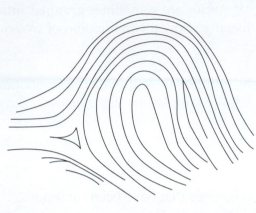

Print #1

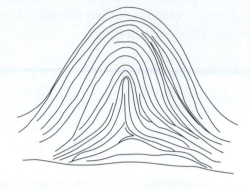

Print #2

X	Characteristic
	Delta
	Whorl
	Simple arch
	Tented arch
	Loop
	Double loop
	Mixed
	Divergence of ridgelines
	Bifurcated ridgelines

X	Characteristic
	Delta
	Whorl
	Simple arch
	Tented arch
	Loop
	Double loop
	Mixed
	Divergence of ridgelines
	Bifurcated ridgelines

FIGURE 18-6. Analysis of Fingerprints

Check your answers with your instructor before you continue.

ACTIVITY 2   THE FINGERPRINT "FORMULA"

A fingerprint formula is a list of **print classifications for each hand.** Each hand has a distinctive fingerprint formula. Before you can analyze crime-scene prints, you must develop some expertise in recognizing the different fingerprint patterns.

1. Work in groups. Get the following supplies: **one fingerprint inkpad per group, magnifying glasses, and some paper towels.**

2. Using the fingerprint inkpad, make a clean set of fingerprints for each member of your team.

    Place the prints on the **Fingerprint File Card** in **Figure 18-7.** To make the prints, place **one side** of each finger on the inkpad. **ROLL** the finger from one side to the other, covering the **FRONT** (**NOT** the tip) with ink.

> ### Note:
> **Be careful with the fingerprint ink. It can stain your clothes.**

3. Repeat the rolling motion to transfer the inked print onto **Figure 18-7** (or a clean sheet of white paper).

    If you press too hard, you'll get a black ink blob with no discernable ridges. Use a light touch to obtain a readable print.

4. To prepare an individual's fingerprint formula, **begin at the thumb and proceed to the little finger.** For example, a hand that has fingers of **arch, arch, whorl, loop, whorl** has a print formula of **a–a–w–l–w.**

    Using your own fingerprints, **determine your fingerprint formula.** Write the fingerprint formula **directly underneath each fingerprint in Figure 18-7.**

Left Hand   Thumb

Right Hand

Thumb

FIGURE 18-7.   Fingerprint File Card

5. **By group discussion, double-check** each person's assessment of their own finger-print formula.

   Hard-to-classify prints may require your group to clarify and add information to the definitions of each category. As the distinctions between print patterns become more clearly defined, it'll be easier for you to assign individual prints to a print type.

Check your answers with your instructor before you continue.

6. Using your revised print type descriptions, **classify the 12 fingerprints in Figure 18-8.**

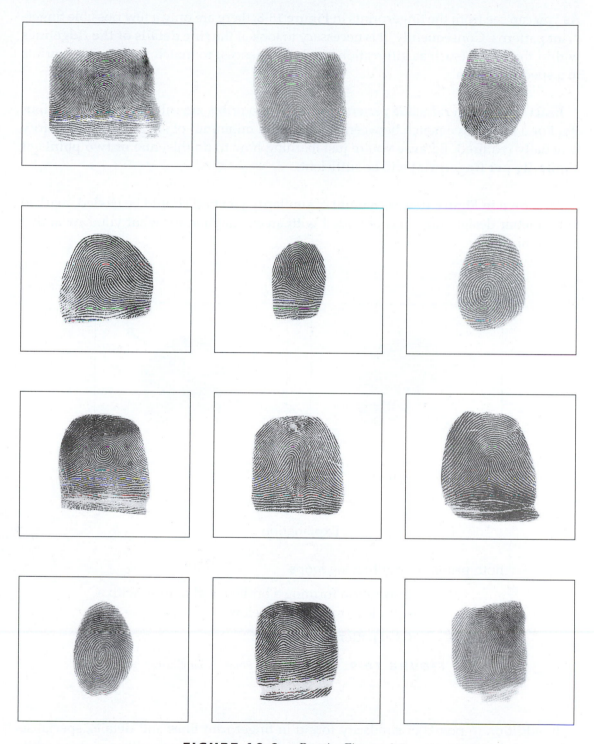

FIGURE 18-8. Practice Fingerprints

 Check your answers with your instructor before you continue.

ACTIVITY 3   DETERMINING POINTS OF SIMILARITY

As you can see from the fingerprints in **Figure 18-8,** there are only a few possible finger-print patterns. Consequently, it is necessary to look at the **fine details of the ridgelines** within the prints (such as bifurcations and divergences) to match sample fingerprints to a specific person.

Exact matches in ridgeline patterns between two prints are called **points of similarity.** For a conclusive match between two prints, a minimum of six points of similarity is usually required. Because you're just learning how to do this, **one or two points of similarity per fingerprint** will be sufficient.

As shown in **Figure 18-9,** each point of similarity you find should be circled, marked with a letter designation, and provided with an explanation for what you saw at that particular point.

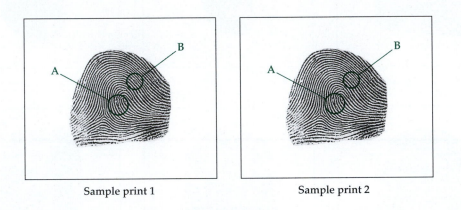

Sample print 1   Sample print 2

Explanation

Both prints are right-facing loops

A   Bifurcation forming a broken oval pattern with a
        complete oval pattern below

B   Oval pattern in ridgeline

FIGURE 18-9.   How to Label Points of Similarity

In addition to points of similarity found in fingerprint lines and details, specialists also check for distinctive scars or marks, which usually show up on the prints as white lines or patterns. Check back to **Figure 18-7,** where you placed your own fingerprints.

1. In **Figure 18-7,** circle and label **two locations** where scars are found within your fingerprint patterns.

✔ Comprehension Check

Mr. Akumba arrives home and discovers that his back porch window has been forced open. He looks around the house and discovers that his new TV is missing. When the police arrive, they dust the scene for fingerprints. A few days later, Ms. Wright is arrested while breaking and entering another house in the neighborhood. You're the fingerprint examiner for the police department. The chief detective has asked you to compare Ms. Wright's fingerprints with those found at Mr. Akumba's house (**Figures 18-10** and **18-11**).

Left Hand   Thumb

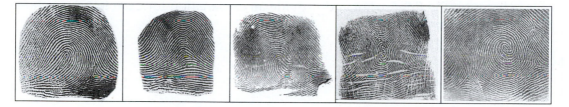

Right Hand

Thumb

FIGURE 18-10.   Fingerprint File Card for Ms. Wright

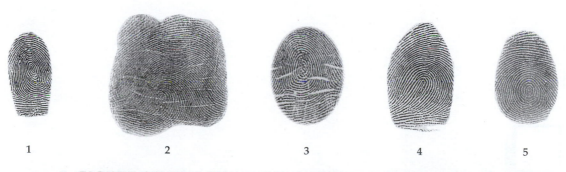

  1   2   3   4   5

FIGURE 18-11.   Unknown Prints Lifted from Mr. Akumba's Window

1. Do any of the prints from the crime scene match those of the suspect? If so, which prints match which fingers? List the matches below.

2. List the **points of similarity** you used to make your match. Mark each point with a letter on the fingerprint and provide an explanation for each letter (as shown in **Figure 18-9**).

3. **Did Ms. Wright do wrong?** Were those prints at Mr. Akumba's house hers? In a few sentences, **summarize the conclusion** you reached on the basis of your fingerprint examination.

Check your answers with your instructor before you continue.

ACTIVITY 4   BLOOD TYPING

Scientists have identified and studied many different human blood groups. Blood groupings are based on the presence of **identifying proteins (also called antigens)** that are incorporated into the cell membranes of all body cells. These molecules are referred to as antigens because they are identified as foreign if placed in the body of a person who doesn't have these molecules.

The most commonly known antigens are those that determine the **A–B–O** and **Rh** blood groups. These blood groups are well known because they are used to type blood for transfusions and organ transplants. The groups include the following:

- **Type A** blood contains antigen **A** and forms antibodies against **B**.

- **Type B** blood contains antigen **B** and forms antibodies against **A**.

- **Type AB** blood contains both antigens **A and B.** Type AB **doesn't form antibodies against either A or B.**

- **Type O** doesn't contain **A or B antigens** but forms antibodies against **both A and B.**

- The **Rh factor** is the name given to another antigen present in red blood cells. People with **Rh+ blood have the antigen,** while those with **Rh– blood don't.**

Naturally occurring **antibodies (defensive proteins)** in the blood cause a serious transfusion reaction called **agglutination** if a person is transfused with blood containing **foreign** antigens of the A–B–O or Rh blood groups. Babies aren't born with these specialized antibodies, but they begin to develop them a few months after birth.

The combination of an antigen and its corresponding antibody produces **agglutination (clumping of blood cells).** The agglutination reaction makes it possible to determine the blood type.

To determine a person's blood type, we'll cause agglutination to happen in the laboratory. Of course, we can't do this test in a person's bloodstream, because that could cause harmful agglutination within the blood vessels. However, we can place the blood samples in the wells of a depression plate and simulate the reaction, so that you can observe how the clumping takes place.

The antibodies that bind to the blood antigens are labeled according to the antigen they bind to. For example, **antibody-A (which we'll refer to as "anti-A") binds to and causes clumping in type A blood (because type A blood contains the matching type A antigen).**

> **Note:**
> If we add anti-A to type A blood, clumping will occur. The addition of
> antibody-B, however, wouldn't cause these cells to clump.

By analyzing the reactions of an unknown blood sample with specific antibodies, and comparing your results to those in **Table 18-1,** the blood type can be determined.

1. Work in groups. Get the following supplies: **a spot plate, a wax pencil, some tooth-picks, a magnifying glass, and dropper bottles of the following: known samples 1 and 2, unknown samples 1 and 2, anti-A serum, anti-B serum, and anti-Rh serum.**

> **Note:**
> These tests are being carried out with **simulated blood, which is nonbiological**
> and nontoxic.

TABLE 18-1
CLUMPING REACTIONS THAT OCCUR IN BLOOD TYPING

BLOOD TYPE	EFFECT OF ADDING ANTIBODY-A	EFFECT OF ADDING ANTIBODY-B
A	Clumps form	No clumps
B	No clumps	Clumps form
AB	Clumps form	Clumps form
O	No clumps	No clumps

2. You'll be testing each blood sample for the presence of the antigens A, B, and Rh.

    With the wax pencil, label **three horizontal rows** of your spot plate **A, B,** and **Rh** as shown in the sample in **Figure 18-12.**

3. You'll be practicing on two blood samples of **known** type **(K1** and **K2).** With these two samples as **controls,** you'll be able to see exactly what the clumping patterns look like for each possibility of A, B, and Rh antigens. **Your instructor will let you know the blood types of the known samples.**

    Fill in the blanks next to K1 and K2 in the first two vertical columns of your spot plate with the **names of the blood types provided by your instructor.**

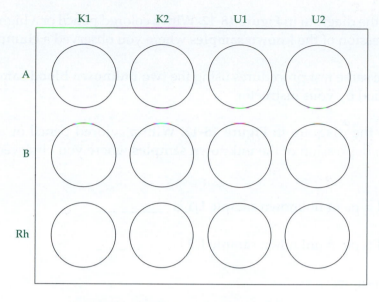

FIGURE 18-12. Sample Spot Plate

4. Place one large drop of **anti-A** solution into **depressions K1 and K2** in the row labeled **A**. Place one large drop of **anti-B** solution into **depressions K1 and K2** in the row labeled **B**.

    Place one large **drop of anti-Rh solution** into **depressions K1 and K2** in the row labeled **Rh**.

5. For each of the three depressions for the known **sample K1,** put a drop of blood from the appropriate dropper bottle.

    Stir each sample with a toothpick. **Use a CLEAN TOOTHPICK for each sample.**

6. For each of the three depressions for known **sample K2,** put a drop of blood from the appropriate dropper bottle.

    Stir each sample with a toothpick. **Use a CLEAN TOOTHPICK for each sample.**

7. Observe the mixtures in the six depressions.

    If the agglutination test is **positive,** you'll see a **collection of small clumps,** or particles, in the depression.

    You may find this easier to see by using a magnifying glass or dissecting microscope.

> ### Note:
> **Don't empty the liquid from the depressions containing the controls.**

8. Return to the diagram in **Figure 18-12.** With a colored pencil or a highlighter, **color** each depression of the **known samples** where you observed a **clumping** reaction.

9. Repeat the same test procedures using the **two unknown blood samples (U1 and U2)** supplied by your instructor.

10. Return to the diagram in **Figure 18-12.** With a colored pencil or a highlighter, **color** each depression of the **unknown samples** where you observed a **clumping** reaction.

11. The blood type of unknown **sample U1** is _____.

    The blood type of unknown **sample U2** is _____.

    Explain your answers.

✓ Comprehension Check

Owing to a clerical error, several samples of blood stored at the local blood bank may be incorrectly labeled as to the blood type. The three tests listed below were conducted on each sample. Use your knowledge of antigen–antibody reactions to sort the samples into their correct blood types.

Test 1   Unknown sample mixed with "anti-A" serum (contains "type A" antibodies)

Test 2   Unknown sample mixed with "anti-B" serum (contains "type B" antibodies)

Test 3   Unknown sample mixed with "anti-Rh" serum (contains antibodies against the Rh protein)

1. **Unknown Sample A:** When tested, agglutination (clumping) of red blood cells occurred in Tests 1, 2, and 3.

   Blood Type: _____

2. **Unknown Sample B:** When tested, no agglutination occurred in any of the three tests.

   Blood Type: _____

3. **Unknown Sample C:** When tested, agglutination occurred in Tests 2 and 3, but not with Test 1.

   Blood Type: _____

4. **In your OWN words,** explain the chemical reaction that occurs when anti-B serum is mixed with

   a. a type A blood sample:

   b. a type B blood sample:

c. an Rh-positive blood sample:

Check your answers with your instructor before you continue.

Note:

Blood typing can be useful; however, it isn't highly informative. Because there are only four blood groups in the A–B–O system, many people share the same type. Even if the blood type of the evidence matches the suspect, it doesn't prove that this blood came from the suspect. It could be from anyone that has the same blood type. For this reason, forensic scientists have moved away from conventional blood typing toward the more specific DNA typing technology.

Although millions of people have the same blood type, there are differences in the frequency of each blood type in various groups. The table below summarizes the distribution of the A–B–O blood types in four major ethnic populations in the United States.

ETHNIC GROUP	TYPE A (%)	TYPE B (%)	TYPE AB (%)	TYPE O (%)
Caucasian	40	11	3	46
African American	27	20	3	50
Hispanic	31	10	2	57
Asian	28	25	7	40

SELF TEST

Fill in the blanks with the choice that is **most appropriate. Answers can be used only once.**

a. ridgeline
b. delta
c. whorl
d. tented arch
e. loop

f. bifurcation
g. divergence
h. fingerprint formula
i. points of similarity
j. sweat glands

1. _____ Complete circle formed by ridgelines between two deltas.

2. _____ Two parallel ridgelines spread apart.

3. _____ Ridgeline divides into two or more branches.

4. _____ Curved ridgelines that enter and exit on the same side of the print.

5. _____ Produce fluids that form latent fingerprints.

6. _____ Set of print classifications for each hand.

7. _____ Exact matches of ridgeline patterns between two fingerprints.

8. _____ Fingerprint pattern in which ridges meet at the center to form a peak.

9. Why do your fingers leave prints when you touch something?

10. Why is it unlikely that identical twins would have identical fingerprints?

For questions 11–13, identify the most likely blood type on the basis of the results of the following antigen–antibody reactions. Explain each answer.

11. When tested, agglutination occurred when exposed to anti-A serum, but not with anti-B or anti-Rh serums.

12. When tested, agglutination occurred when exposed to anti-A serum, and also with anti-B serum. No agglutination occurred with anti-Rh serum.

13. When tested, agglutination didn't occur when exposed to both anti-A and anti-B serums. Agglutination did occur with anti-Rh serum.

Biotechnology: DNA Analysis

Objectives

After completing this exercise, you should be able to:

- discuss the differences between genes and noncoding regions of DNA
- explain the role of noncoding DNA regions in producing a DNA profile
- draw conclusions that are supported by analysis of RFLP and STR profiles
- interpret DNA data presented in the form of tables, charts, or graphs
- explain how STR analysis can be used to determine paternity
- apply your knowledge of DNA fingerprinting to real-life situations

CONTENT FOCUS

Most of you are aware of the events of September 11, 2001, when there was a terrorist attack against the World Trade Center in New York City. More than 2,700 people died during the collapse of the two World Trade Center buildings. Because it was impossible to identify most of the victims from their remains, DNA testing was used to identify bone and tissue fragments found at the site.

How does DNA technology work? Half of your 46 chromosomes were inherited from your mother and the other half from your father. Consequently, **everyone has a unique DNA pattern** (except for identical twins). DNA profiling technology is complex because 99.9% of everyone's DNA is the same. So individuals are separated based on the remaining 0.1% of the DNA code.

DNA is located in the nucleus of most body cells. Therefore, DNA can be extracted from any tissue. Common sources include blood, hair roots, teeth, semen, saliva (which contains epithelial cells), and other body tissues. Tissue samples containing DNA can be used for identification in criminal cases; in paternity suits; and in the case of the World Trade Center, when visual identification wasn't possible. The technique used to make these genetic comparisons produces a **DNA profile** (also called a **DNA fingerprint**).

When a DNA profile is collected as evidence, law enforcement agencies can use a software program called **CODIS** (Combined DNA Index System) that operates databases of DNA profiles for convicted criminals, evidence from unsolved crimes, and missing persons. When evidence from one crime scene is compared with evidence from another using CODIS, it's possible to discover if the same person was involved in previous crimes in other locations.

DNA can even be found on evidence that may be decades old. Therefore, "cold cases" that were never solved may be reopened. In the past several years, some convicted criminals have been released based on DNA evidence.

ACTIVITY 1   SOURCES OF DNA EVIDENCE

Only a few cells or a small fluid sample is needed to construct a DNA profile. The following is a partial list of some common pieces of physical evidence that might be collected at a crime scene.

Following the example in **Table 19-1,** list several sources of DNA evidence from items collected at a crime scene.

TABLE 19-1 **SOURCES OF DNA EVIDENCE**	
EVIDENCE COLLECTED AT THE CRIME SCENE	SOURCES OF DNA
eyeglasses	sweat, skin cells, hair
baseball bat	
toothpick	
sheets and pillows	
used kitchen spoon	
men's or women's underwear	
sealed, stamped envelope	

Check your answers with your instructor before you continue.

ACTIVITY 2   SOLVING A CRIME USING RFLP FINGERPRINTS

An early form of forensic DNA analysis used **restriction enzymes** (enzymes that cut the DNA molecule at specific sites) to produce fragments of DNA for testing. This method is called **RFLP** analysis (restriction fragment length polymorphism). Because RFLP analysis requires relatively large amounts of DNA, it isn't often used any more. There are newer, more efficient DNA analysis techniques. However, for an understanding of the principles underlying DNA analysis, it's still worthwhile to examine how this technique is used to match DNA samples.

DNA testing produces an image with a pattern of bands for several DNA locations (see **Figure 19-1**). The bands are analyzed to create a **DNA profile** for an individual DNA sample. The separate samples can be used to compare DNA recovered from a crime scene and DNA from a suspect. Each separate DNA sample is represented by a **vertical column** of DNA bands.

For RFLP testing, **several different regions** (called **loci**) of DNA are analyzed. This results in the banded pattern you can observe in each DNA profile in **Figure 19-1**. If **any** of the loci **don't match** between an evidence sample and a known individual, **the person**

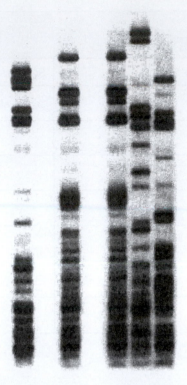

FIGURE 19-1.   Sample DNA Fingerprints

is excluded as a possible source of the evidence DNA. If, however, **ALL** of the loci in the evidence sample exhibit the exact same pattern as the known sample, that person cannot be eliminated as a suspect.

✔ Comprehension Check

1. How many separate DNA samples are shown in **Figure 19-1?** _____

2. Do any of the samples match? Explain your answer.

Check your answers with your instructor before you continue.

The banding pattern you observed in **Figure 19-1** is difficult to interpret with the naked eye so the DNA profiles are often shown in chart format (as in **Figure 19-2**). **Each separate DNA sample is presented as one vertical column. The horizontal bands that represent the DNA loci are shown within that vertical column.**

Use the RFLP chart in **Figure 19-2** to solve the following crime:

On January 1, at about 2:00 a.m., the police responded to a report of gunshots at the 600 block of Knorr Street. Arriving at the scene, the officers found a parked Dodge pickup. In the front seat was a deceased male with several gunshot wounds. Shoe marks were visible in a pool of blood outside the truck, but they weren't clear enough to identify the type of footwear that made them. Homicide detectives have narrowed the field to three suspects.

Suspect A was arrested near the scene. He was wearing black work boots that appeared to have dried blood on the soles. When interviewed, Suspect A said he had no idea there was blood on his shoes and no clue where it came from.

At the home of **Suspect B,** investigators recovered a pair of tennis shoes that also appeared to have dried blood on the soles. Suspect B says she had a severe nosebleed, which might have resulted in blood on her shoes.

Suspect C was the roommate of the deceased. When officers went to the apartment to notify him of the death, Suspect C was packing a suitcase. He was wearing high-top sneakers that appeared to have dried blood on the soles and sides. When questioned about the blood and the murder, Suspect C had no response.

All the samples were tested and identified as human blood and submitted for DNA analysis. You're the forensic scientist assigned to the case. **Figure 19-2** shows the DNA profiles obtained from the blood samples.

Note:

When trying to determine whether two DNA profiles are a match, remember that 100% of the bands have to be the same. If even one band doesn't match, the profiles are not from the same person.

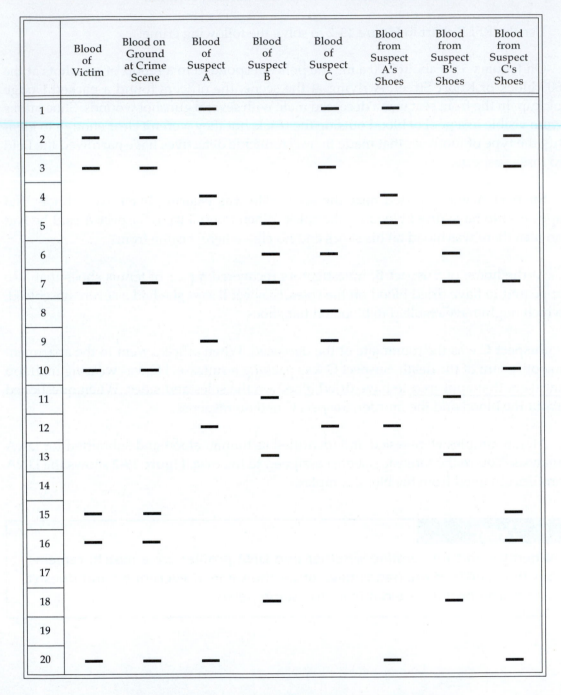

FIGURE 19-2.   DNA Profiles—Knorr Street Homicide

✓ Comprehension Check

1. Did any of the profiles match the blood from **Suspect A's shoes?** If so, list them:

2. Did any of the profiles match the blood from **Suspect B's shoes?** If so, list them:

3. Did any of the profiles match the blood from **Suspect C's shoes?** If so, list them:

4. Did any of the profiles match the blood from the deceased male? If so, list them:

5. Did any of the profiles match the blood found on the ground outside the truck? If so, list them:

6. Do these results make any individual(s) more likely suspects than the others? Explain your answer, citing facts from the DNA analysis.

7. List several other types of evidence that would be helpful to be sure you've found the murderer.

Check your answers with your instructor before you continue.

ACTIVITY 3   THE USE OF PCR FOR STR ANALYSIS

The amount of DNA collected in some situations is very small and normally wouldn't be enough for analysis (such as the back of a licked envelope). To provide enough DNA for analysis, regions of interest are copied (amplified) with a technique known as **polymerase chain reaction** (**PCR**). PCR produces millions of exact copies of the DNA sample.

How small a sample would be enough for DNA analysis if amplified with PCR? Consider **Figure 19-3,** which compares the size of an average shirt button with the size of a suitable DNA sample.

Various traits are produced on the basis of your genetic code (nuclear DNA). However, not all of a person's DNA codes for the synthesis of proteins. Regions that don't code for proteins are referred to as "**noncoding**" **DNA.** Noncoding DNA varies a great deal among individuals, and when specific DNA regions are studied, scientists can use this information to establish human identity.

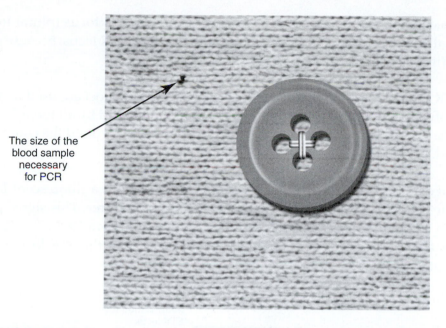

The size of the
blood sample
necessary
for PCR

FIGURE 19-3. Tissue Sample Suitable for Analysis Using PCR Amplification

The pattern of noncoding regions is unique and always has the same repeating pattern of nucleotides (for example, **C A T**). The length of the repeating sequence of bases is variable and the number of times these nucleotides are repeated is also highly variable among people.

As you can see in **Figure 19-4,** Person A has **nine** repeating **C A T** segments, while Person B has only **five.** Some people have dozens of repeats of the same pattern. Each person has a distinct **DNA profile** (or fingerprint).

The repeating sections of noncoding DNA are called **short tandem repeats (STRs)** in forensic DNA testing. As with all of your DNA, noncoding DNA is inherited—half your DNA comes from each parent.

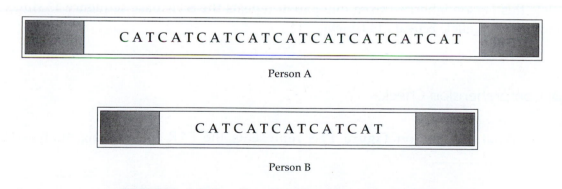

CATCATCATCATCATCATCATCAT

Person A

CATCATCATCATCAT

Person B

FIGURE 19-4. Repeating Noncoding DNA Segments

Each section of noncoding DNA that's analyzed is called a **locus** (plural **loci**). Each individual would have two copies (called **alleles**) for each STR locus, because each person inherits one allele from each parent.

An STR profile doesn't consist of only one locus. Multiple loci are used to develop a DNA profile. In the United States, the same STR loci are tested by all forensic laboratories. By having all labs perform the same tests, results from around the country can be compared.

To make an STR profile, the DNA sample is tagged with a fluorescent label. The label is detected by a laser, which sends a signal to a computer. This signal produces a peak on a computer graph (similar to the example in **Figure 19-5**). The **letters and numbers above the peaks** are abbreviations that represent the STR loci used in the CODIS database.

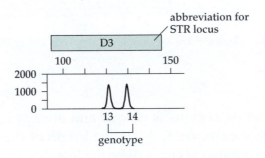

FIGURE 19-5.   Computer Printout
of Sample Alleles at One STR Locus

Note that there are **numbers below each peak.** These numbers are used to designate the person's **genotype** (his or her two alleles) at that specific locus.

If you see only **one peak** instead of the usual two, that individual has inherited **two alleles that are alike** at that locus. The term used when a person inherits two copies of the same allele is **homozygous.** Inheritance of two different alleles is called **heterozygous.**

In the sample locus in **Figure 19-5,** the genotype would be **13, 14** (meaning that the allele this person inherited from **one parent repeats the STR base sequence 13 times** and the other allele **repeats 14 times**).

☑ Comprehension Check

1.   What number on **Figure 19-5** represents the STR locus being analyzed?
     _____

2.   **(Circle one answer.)** The locus shown in Figure 19-5 is **homozygous / heterozygous.** Explain your answer.

Check your answers with your instructor before you continue.

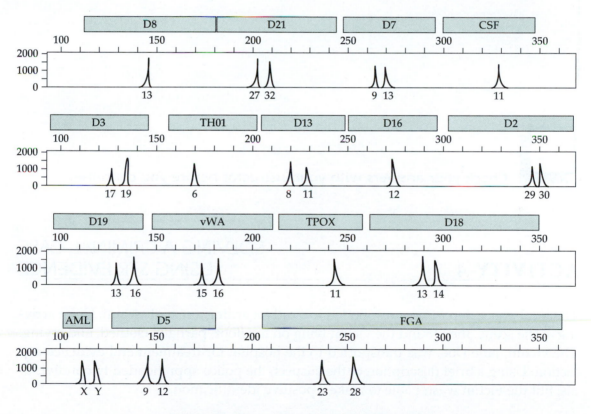

FIGURE 19-6.   Example of an STR Profile

A complete STR profile is shown in **Figure 19-6.** The forensic community has adopted a set of core STR loci for analysis. The locus for gender identification is abbreviated **AML.** Males are designated XY and females as XX.

✔ Comprehension Check

1. **(Circle one answer.)** This DNA profile was taken from a **male / female.**

2. How many STR loci in the profile are homozygous? _____

    How many are heterozygous? _____

3. What do the numbers **9** and **13** represent in the D7 locus? Explain your answer.

Check your answers with your instructor before you continue.

ACTIVITY 4

SOLVING A CRIMINAL CASE USING STR EVIDENCE

The local police department responded to a call regarding a sexual assault. Upon arriving at the scene, they found the victim lying on the floor, partially clothed, and crying. The victim, Jane Doe, was transported to the hospital for treatment and evidence collection. Using a brief description of the suspect, the police apprehended two individuals, but the victim wasn't able to make a positive identification.

At the hospital, **two evidence samples** were collected from the body of the victim. The evidence samples **tested positive for semen** and were sent to the lab for STR analysis. In addition, **known blood samples from all individuals** were obtained and submitted for STR analysis.

1. Interpret the STR profiles in **Figures 19-7** through **19-11** on the following pages and **record the genotypes** for each locus in **Table 19-2**.

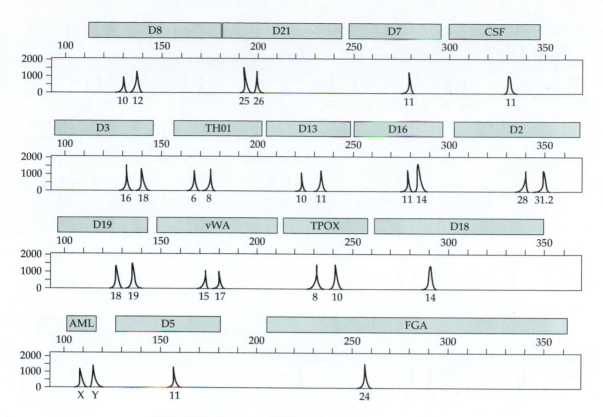

FIGURE 19-7. STR Profile of Semen Sample One

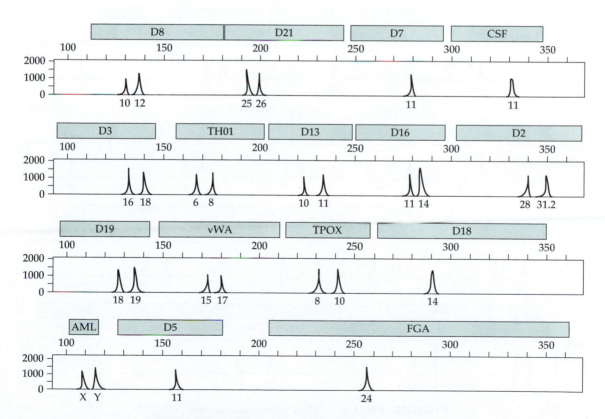

FIGURE 19-8. STR Profile of Semen Sample Two

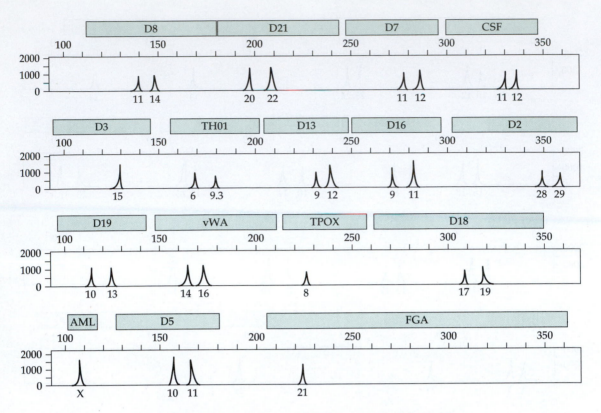

FIGURE 19-9. STR Profile from the Victim

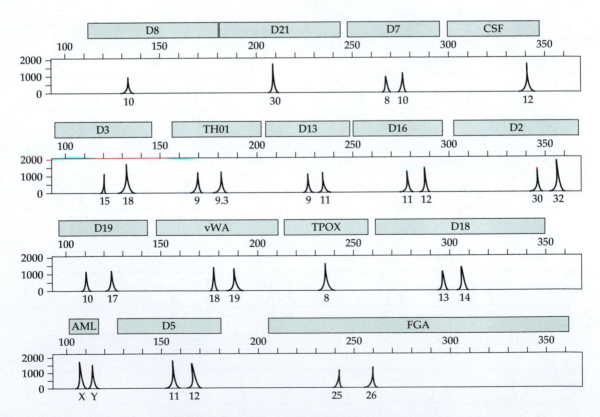

FIGURE 19-10. STR Profile from Suspect One

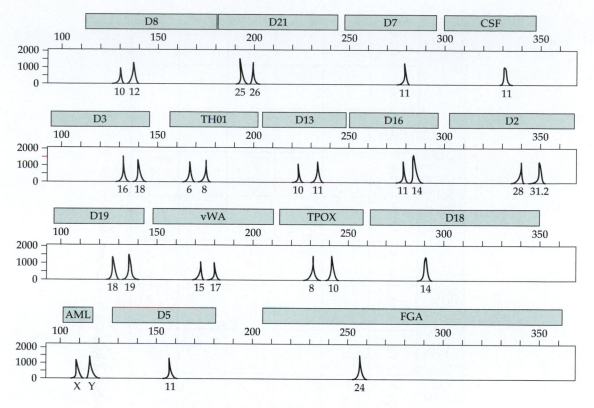

FIGURE 19-11. STR Profile from Suspect Two

TABLE 19-2
STR PROFILES FROM SEXUAL ASSAULT EVIDENCE

DNA Samples	D8	D21	D7	CSF	D3	TH01	D13	D16	D2	D19	vWA	TPOX	D18	AML	D5	FGA
Semen Sample 1																
Semen Sample 2																
Victim																
Suspect 1																
Suspect 2																

Note:

When trying to determine whether two STR profiles are a match, remember that the alleles at 100% of the loci have to be the same. If even one locus doesn't match, the profiles are not from the same person.

✓ Comprehension Check

1. According to the results of your analysis, what conclusions can you draw about this case? **Be specific. Support your answer with facts** from the STR profiles.

2. The detective investigating this case asks you the following question. Please respond.

   "If the DNA profile from the evidence matches the suspect, he must be guilty of sexual assault, right?"

3. **(Choose one answer.)** When a blood sample is submitted for STR analysis, the DNA tested is taken from **red blood cells / white blood cells / platelets. Explain** your answer.

4. What results would you expect from a comparison of STR profiles taken from the following? **Explain** your answers.

    a. Identical twins:

    b. Fraternal twins:

Check your answers with your instructor before you continue.

ACTIVITY 5

DECIDING A PATERNITY SUIT
INVOLVING STR ANALYSIS

STR profiles can also be used to help identify a child's biological parents. As you recall from the lessons on meiosis, a child inherits half of its genetic information from each parent. This makes matching of DNA profiles more challenging.

If a child has alleles in his/her STR profile that don't match those of the mother, these alleles must have been inherited from the father.

A young woman claims that her two-year-old daughter is the child of one of the members of the hit rock group "First Impressions." During the time of conception, she was dating all three of the band members.

In her sworn testimony, the mother states, "The father is one of the band members. They were the people I spent time with during that whole year. The problem is, I don't know which musician is the biological father of my baby."

None of the members of the band have admitted to fathering the child. During the investigation, all members of the rock band agreed to provide DNA samples for analysis. You're the forensic scientist appointed by the court to analyze the DNA evidence.

The profiles for all parties involved are summarized in the STR table **(Table 19-3).**

1. Compare the child's STR profile with the mother's. In the child's profile, **cross out** any alleles that **are the same as the mother's** (as shown in green ink in **Table 19-3).**

2. Next, compare the alleles of each of the potential fathers with the child's profile. **Circle all matching alleles.**

3. On the basis of your analysis of the STR profiles, which of the men could be the father of the child? **Explain** your answer, using facts from the STR analysis.

Check your answers with your instructor before you continue.

TABLE 19-3
STR PROFILES FOR PATERNITY SUIT

DNA SAMPLES	D8	D21	D7	CSF	D3	TH01	D13	D16	D2	D19	vWA	TPOX	D18	AML	D5	FGA
Child	10,19	29,31	6,9	13,14	13,18	9,11	13,13	11,14	28,30	14,17	12,12	7,11	10,12	XX	7,16	20,25
Jenny Smith	9,19	29,22	6,6	6,14	12,18	10,11	13,14	8,11	28,32	14,30	12,15	7,13	12,19	XX	12,16	20,20
Peter Niles	9,10	18,31	9,11	12,14	13,13	5,7	8,9	9,13	31,32	32,33	11,12	8,9	9,10	XY	10,12	22,25
Phillip Cruze	10,19	19,31	9,15	11,13	12,13	9,10	11,13	14,16	30,30	8,17	12,20	11,13	10,19	XY	7,9	25,29
Mike Miller	10,15	18,31	9,6	9,13	13,19	7,11	15,15	11,16	31,31	15,30	12,12	8,12	14,15	XY	13,16	24,26

SELF TEST

1. What is meant by "noncoding" DNA regions?

2. How can noncoding DNA regions be used to determine a person's identity? **Be specific.**

3. How is comparing STR profiles to determine paternity different from trying to match a suspect to a blood sample?

4. You're a park ranger at Yellowstone National Park. You've discovered the internal organs of a deer just inside the park boundaries. Because hunting in the park is illegal, you notify the local police to be on the alert for a recently killed deer. The police spot a truck carrying the carcass of a large deer a few miles outside the park. They suspect this is the deer that was killed inside the park boundaries, but they need some evidence linking this deer to the remains found in the park.

   What test(s) could determine whether the deer in the truck was killed inside Yellowstone Park? **Explain** your answer.

5. On Saturday, September 28, a young boy was bitten by a large black dog on the corner of Fifth Avenue. Investigating the situation further, police officers learned that a neighbor has a large black dog that often barks ferociously at people walking down the street. The owner maintained that the dog made a lot of noise, but was basically friendly. The officers, however, discovered a substance, which appeared to be blood, on the dog's collar. The owner states that she cut her hand on the dog collar and that the blood is hers.

    The boy's family decides to sue the dog's owner. They wish to be compensated financially and also have the dog destroyed. However, the dog's owner is adamant that her dog never left her fenced yard on the date in question. You're a forensic scientist for a private firm that has been asked to evaluate the case for the civil trial. All samples have been tested and found to be human blood. Interpret the results of the DNA testing as presented in **Figures 19-12** through **19-14.**

    Summarize your findings in this investigation. **Support your statements with evidence** from the DNA fingerprints.

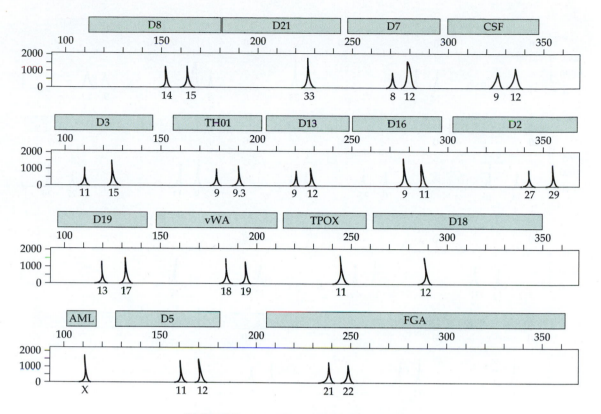

FIGURE 19-12. Dog Owner's Blood

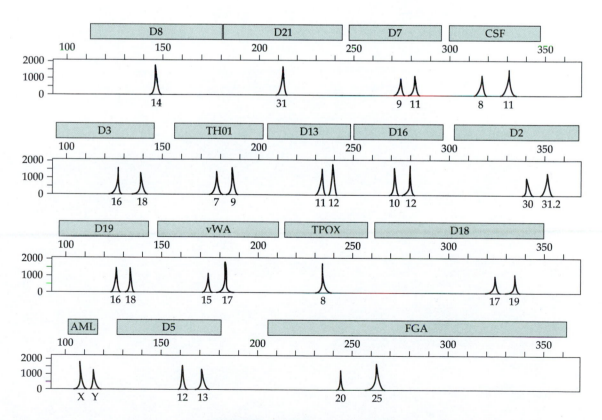

FIGURE 19-13. Boy's Blood

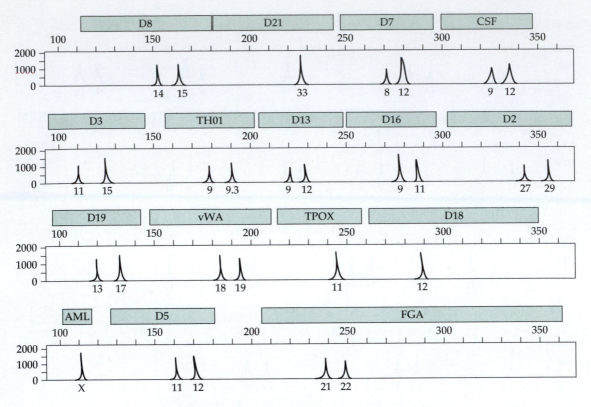

FIGURE 19-14. Blood from Dog's Collar

Using Biotechnology to Assess Ecosystem Damage

Objectives

After completing this exercise, you should be able to:

- explain the effects of environmental factors on bioluminescence and give examples
- correctly prepare and use serial dilutions
- demonstrate awareness of factors that can affect the accuracy of experimental results
- explain the utility of serial dilutions to analyze the effects of commonly used chemicals on ecosystems
- use experimental results to rank products in terms of their potential to cause harm to human health or the environment
- draw graphs that present data clearly and accurately
- interpret data in tables, charts, and graphs
- apply your knowledge of chemical toxicity to real-life situations

CONTENT FOCUS

Many types of harmful compounds are released into the environment by industry, by agriculture, and even from our own households. It's important not only to detect harmful materials but also to predict their effects on the environment. This is difficult because pollutants are often released in **small amounts, which rapidly disperse** into the soil or water. It's also important to identify **which types** of chemicals are present in the environment, and whether these chemicals are **harmful** to living organisms.

To address this problem, Dr. Kenneth Thomulka, a scientist from Philadelphia, developed an inexpensive, convenient field-testing method to discover whether toxic chemicals are present in the environment.

The method uses harmless marine bacteria as **biological indicators** for the presence of toxic chemicals.

ACTIVITY 1   EXPLORING BIOLUMINESCENCE

The ability of an organism to produce light is called **bioluminescence.** Bioluminescence is exhibited in a variety of organisms, from bacteria to fireflies in your backyard. It is the only light source for marine organisms living deep in the ocean. Like fireflies, some deep-sea fishes use their lights as signals to find mates. Others, like the deep-sea anglerfish, wave glowing lures to attract smaller fish as prey.

Many bioluminescent organisms, including bacteria, have the enzyme **luciferase.** Luciferase and **oxygen** are needed for the complex chemical reactions used to produce light.

Note:
Read these instructions COMPLETELY BEFORE the lights go out!

---

1. Work in groups. Get the following supplies: **a test tube rack and one 15-ml screw-capped test tube containing a sample of the bacterial culture.**

2. Place the test tube in the rack and set it on your laboratory table. **Unscrew and remove the lid.** Leave the test tube **undisturbed** for 15 minutes.

Caution:
At times during this laboratory period, the room will be completely dark. Make sure the aisles and working areas around your table are clear.

---

3. To demonstrate the production of light, the class will observe a flask containing **bioluminescent marine bacteria.** With the **lights out,** observe what happens when the instructor swirls the flask containing the bacteria. **Record** your observations.

4. **Continue to observe** the flask for several minutes as it sits undisturbed on the laboratory counter. **Record** your observations.

5. **Without moving or disturbing the test tube rack,** check the tube of bacteria on your laboratory table. Observe the tube **very carefully.**

   Where is the light level **most** intense? _____

   Where is the light level **least** intense? _____

   Using the information presented in the **introduction** to this exercise, which mentions **two** substances needed for bioluminescence to occur, **explain** your observations.

6. **Rank the level of light produced** by the bacteria on a scale of **0 (no light) to 4 (most light).**

   Top of the tube _____

   Middle of the tube _____

   Bottom of the tube _____

7. **Shake** the test tube **gently** and observe the results. **Describe** what you see.

8. After shaking the test tube, **rank the level of light produced** by the bacteria on a scale of **0 (no light) to 4 (most light).** _____

   Why did the change in light level occur?

Bacterial test kits are commercially available for detecting pollutants in aquatic ecosystems, such as lakes, streams, and oceans. Manufacturers of household cleaners, shampoos, and cosmetics have started using bacterial testing to replace the more traditional, but highly controversial, tests done with rabbits, rats, and mice.

   In the following experiments, you'll use bioluminescent bacteria to evaluate the toxicity of common household products. You'll try to determine whether any of these household products can be damaging to the environment if they **aren't used and disposed of properly.**

   Any environmental condition that's harmful to the bacteria's health will cause a decrease in light production.

   The point of this experiment ISN'T to see how well you can kill bacteria. These bacteria are not harmful. They are valuable and necessary for the environment. The point is to find out how you can PREVENT household products from harming organisms in the environment, including bacteria.

ACTIVITY 2

1. Work in groups. Get the following supplies: **a small flashlight with a red filter, one small beaker, one graduated 10-ml pipette with manual dispenser, and one graduated 1-ml pipette with manual dispenser.**

    You'll also need the following: **six large test tubes in a rack, six small test tubes in a rack, scissors, masking tape, a dispenser bottle filled with 3% saline solution, a tray for used glassware, and a container for waste fluids.**

2. Fill the small beaker half full with **tap water.** Attach the manual dispenser to the end of a **graduated 10-ml pipette.**

    Practice filling and dispensing fluid from the pipette following the directions given by your instructor.

3. Now that you're an expert pipette user, you are ready to set up your experiment.

    With masking tape, label the **large** test tubes **D1** through **D6.**

    Make sure you place the tape labels as **close as possible to the top of each tube.**

4. The following household products are available for testing:

drain cleaner	automobile antifreeze
mouthwash	daily shower cleaner
household cleaner	herbicide (weed killer)
toilet-bowl cleaner	automobile wheel cleaner

Caution:

Some of these products may be harmful to your skin or eyes. Be careful not to spill any on your hands as you measure. If a spill occurs, DON'T touch your face or eyes. Wash your hands thoroughly before proceeding.

---

5. Your instructor will assign a household product to your group.

    Obtain a small container filled with your assigned household product.

    Enter the name of the product your group will be using: _____

ACTIVITY 3   MAKING ACCURATE SERIAL DILUTIONS

1. You'll **dilute** the chemical with various amounts of saline to test the reaction of the bacteria to **different strengths** of this product.

    The dilutions will produce a series of proportionally weaker solutions, commonly referred to as **serial dilutions.** In this manner, you can determine what dilution (concentration) of this chemical is harmful to the bacteria.

    In this experiment, we'll compare products to see which are more harmful than others, and we'll base our assessment of toxicity on the level of dilution necessary to dispose of the chemical safely.

    Rather than just diluting randomly to make this assessment, it's more useful to dilute the chemical in **measured steps (making a serial dilution).** By exposing bacteria to different concentrations (dilution levels) of the various household products, you can determine the exact level of dilution that will make the product harmless.

    Because these are marine bacteria, you'll make your dilutions with **3% saline** (a mild salt solution), which simulates the water in the bacteria's natural environment.

2. The dilution levels you'll use are listed in **Table 20-1.**

T A B L E   2 0 - 1
DILUTION SEQUENCE

TEST TUBE NUMBER	DILUTION	PERCENT CONCENTRATION OF PRODUCT
D1	Full strength	100
D2	1:10	10
D3	1:100	1
D4	1:1,000	0.1
D5	1:10,000	0.01
D6	No product added	0

> ### Hint:
> **To obtain valid results, you must be very careful to make accurate measurements.**

3. Using the **10-ml graduated pipette,** transfer **9 ml** of your assigned household product into the first **large** test tube (marked **D1**). This will be the **full-strength** tube.

Note:

Place all used pipettes and glassware into the container provided.

---

4. Using a **1-ml graduated pipette,** transfer **1 ml** of household product **from the container with the original sample** into the next large test tube (marked **D2**).

5. To the **same** tube **(D2),** add **9 ml of saline** solution from the stock bottle. **To dispense the saline,**

    - hold the test tube under the spout
    - **slowly** raise the pump handle as far as it will go
    - **gently** push the handle down to dispense

    Without spilling the contents, **shake the test tube gently** to mix the solution. This will be the **1:10 dilution.**

6. **From test tube D2, remove 1 ml** of solution and transfer it to test tube **D3.**

    Following the instructions above, add **9 ml of saline** to tube **D3.**

    Shake gently to mix the contents. This will be the **1:100 dilution.**

7. **From test tube D3, remove 1 ml** of solution and transfer it to test tube **D4.**

    Following the instructions above, add **9 ml of saline** to tube **D4.**

    Shake gently to mix the contents. This will be the **1:1,000 dilution.**

8. **From test tube D4, remove 1 ml** of solution and transfer it to test tube **D5.**

    Following the instructions above, add **9 ml of saline** to tube **D5.**

    Shake gently to mix the contents. This will be the **1:10,000 dilution.**

    Remove 1 ml of solution from test tube **D5** and dispose of it in the **waste container.**

9. To test tube **D6,** add **only 9 ml of saline.** No household product will be added to this tube.

✓ Comprehension Check

1. Which test tube contains the **weakest** sample of household product? _____

2. Which test tube contains the **strongest** sample? _____

3. Which test tube is the **control?** _____

4. Why is a control tube needed in this experiment?

5. Why was 1 ml of liquid **removed from test tube D5?**

6. **In your own words,** explain the **purpose of making a serial dilution** for this experiment.

Check your answers with your instructor before you continue.

ACTIVITY 4   PREPARING BACTERIAL SAMPLES

1. Get the following supplies: **one clean 1-ml pipette.**

    Using the tube of bacterial culture you looked at during Activity 1, transfer **1 ml of bacteria** to each of the **six small test tubes.**

2. To **each** of the six small test tubes, add **9 ml of saline** solution and set them aside.

> ### Wait!
>
> **All laboratory groups must start the next part of the experiment together! While you're waiting, read the instructions for Activity 5 COMPLETELY.**

ACTIVITY 5   TESTING THE TOXICITY OF YOUR CHEMICAL

1. **Record the time** when you begin this procedure. _____

2. **When instructed, carefully empty one small test tube of bacteria into each of the large dilution tubes.**

    Without spilling, **shake each tube gently** to mix the contents.

    Place the empty "bacteria" tubes into the **waste container.**

    It is important to remove these empty tubes from your workspace because the light pollution from the bacteria remaining in the empty tubes will interfere with your observations of the light levels in your experimental tubes.

3. The bacteria/chemical combination must incubate for **15 minutes.**

    Record the time when the incubation period will be completed. _____

4. When the incubation period is completed, observe the bacteria and evaluate the level of bioluminescence present.

Caution:

The room will be COMPLETELY DARK for this activity. Make sure your test tube rack, paper, pen, and flashlight are conveniently positioned before the lights go out!

It will take several minutes for your eyes to become adapted to the dark. To avoid accidents during this period, don't move around.

The flashlight will be turned on only briefly to record your results!

---

5. **Rank the light** produced by your bacteria according to the **0 through 4 scale** you used in Activity 1 and record your results in **Table 20-2**.

TABLE 20-2

LIGHT INTENSITY OF BACTERIA EXPOSED TO SERIAL DILUTIONS OF ONE HOUSEHOLD PRODUCT (0–4 SCALE)

TEST TUBE NUMBER	DILUTION	RANKING (0–4)
D1	Full strength	
D2	1:10	
D3	1:100	
D4	1:1,000	
D5	1:10,000	
D6	No product added	

6. **Record** the results for your group on the **master chart** at the front of the room.

   From the master chart, copy the results for the entire class into **Table 20-3.**

TABLE 20-3

LIGHT INTENSITY OF BACTERIA EXPOSED TO SERIAL DILUTIONS OF ALL EIGHT HOUSEHOLD PRODUCTS (0–4 SCALE)

TUBE #	DRAIN CLEANER	MOUTH-WASH	HOUSE-HOLD CLEANER	ANTIFREEZE	SHOWER CLEANER	WEED KILLER	TOILET-BOWL CLEANER	AUTO WHEEL CLEANER
D1 (full strength)								
D2 (1:10)								
D3 (1:100)								
D4 (1:1,000)								
D5 (1:10,000)								
D6 (control)								

7. Using the data from **Table 20-3,** make **eight bar graphs** that show the results of each experiment in **Figure 20-1.**

   Label each graph with the name of the household product that was tested.

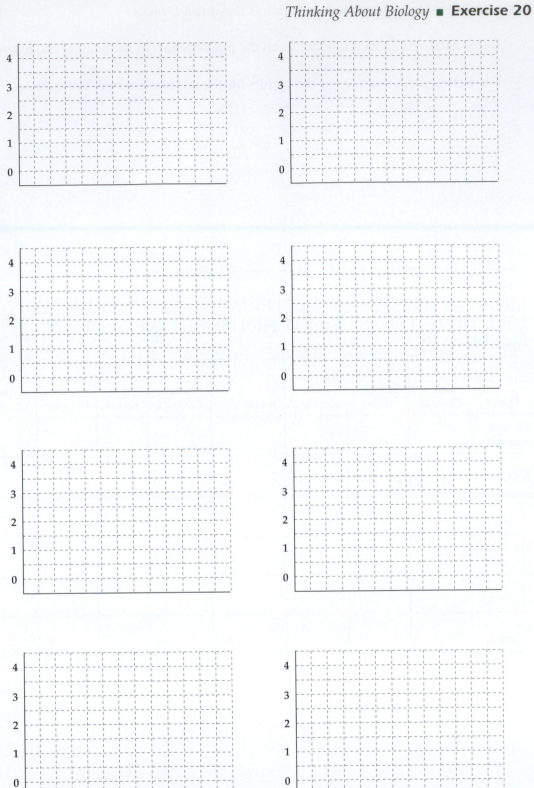

FIGURE 20-1. Comparison of Test Results for Eight Household Products

Check your graphs with your instructor before you continue.

✓ Comprehension Check

1. Which household product was the **most toxic?** _____

   Why did you conclude that this product was the most toxic? In your explanation, **include facts** collected during the bacteria experiments.

2. What part(s) of the food chain might be affected if this product was disposed of improperly in the environment? **Explain** your answer.

3. Imagine this situation. A type of drain cleaner is found to be harmless to bacteria if it is diluted in a 1:10,000 ratio. A 1:10,000 dilution is the same as **1 ml of chemical diluted by 10 liters of water.** The drain cleaner container holds 200 ml of chemical. If the directions tell you to pour the entire contents of the container down the drain, how **many liters of water** would be needed to dilute this chemical to safe levels?_____ **liters　Show your work.**

4. In an experiment testing the toxicity of a car wash product, bacteria exposed to the product at a **1:1,000 dilution** showed a light level of **two.** Bacteria exposed to a **1:10,000 dilution** of the same product showed a light level of **three.**

   In order to dispose of this product safely, is a 1:10,000 dilution sufficient? **Explain** your answer.

5. What would you conclude if **two of the tested products** showed **NO biolumi-nescence** at a **1:1,000 dilution? Explain** your answer.

6. What would you conclude from your experiment if **two** of the tested household products showed the **same light intensity at a 1:10,000 dilution? Explain** your answer.

7. **Challenge Question!** Can the bioluminescence of bacteria exposed to a **1:10,000 dilution** of a chemical ever equal the amount of bioluminescence in the **control tube? Explain** your answer.

Check your answers with your instructor before you continue.

SELF TEST

Figure 20-2 contains the results of a bioluminescent bacteria assay performed with two lawn-care products that frequently find their way into sewers, streams, lakes, and ponds.

1. At which dilution level was **Product A** most toxic? _____

    What is the **lowest** dilution level at which **Product A** was harmless? _____

2. At which dilution level on the graph is **Product B** harmless? _____ **Explain** your answer.

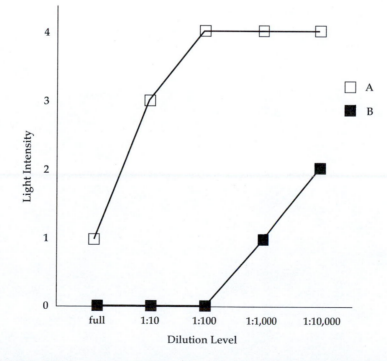

FIGURE 20-2.   Bioluminescent Bacterial Assay of Two Lawn-Care Products

3. If these two lawn-care products do the same job, which product is less harmful to the environment? **Explain** your answer.

Ecosystems

Objectives

After completing this exercise, you should be able to:

- diagram and label a food chain, placing producers and each consumer level in their appropriate positions
- discuss the methods by which nutrients are recycled in ecosystems
- discuss the pathways of energy transfer through food webs and the efficiency of the transfer
- give examples of several factors that can affect the populations at various levels of the food chain
- explain how natural factors and human activities can affect an entire ecosystem

CONTENT FOCUS

An ecosystem is a biological community of plants, animals, and microorganisms (referred to as the **biotic** components of the ecosystem). The biotic communities within a particular ecosystem depend on the physical (**abiotic**) factors present in that location. Some examples of abiotic factors are temperature, availability of water, sunlight, mineral nutrients, and so on. So, an **ecosystem** is defined as the biological community plus the abiotic factors with which those organisms interact. Of course, different ecosystems around the world contain different varieties of organisms. This is due to the large differences in abiotic factors present in different geographic areas.

The energy that makes animal life possible is obtained from **producers.** Producers obtain their energy from **abiotic sources** such as sunlight. The process of taking energy from sunlight is called **photosynthesis.** Examples of producers include plants, trees, algae, and several types of bacteria. **Consumers** must feed on other organisms to obtain the energy to survive. Transfer of energy from producers to consumers is repeated in ecosystems all over the world. Even though deserts and rain forests have different types of producers and consumers, the pattern remains the same.

Decomposers, such as bacteria and fungi, are special types of consumers that feed on the remains of other organisms. During this process, the decomposers receive energy from the food they eat, and **excess minerals and other nutrients are released** to the soil or water to be **recycled** and become available to producers.

Note:

Many different types of bacteria are found in ecosystems. Some are decomposers, some producers, and some parasites that cause diseases. So, it's important to determine the role of each particular type of bacteria before you decide how to classify them in your ecosystem.

To make complex ecosystems easier to study, the producers, consumers, and decomposers in an ecosystem are categorized into food chains. The cycle in **Figure 21-1** shows how energy and nutrients are transferred from producers through consumers and decomposers in a typical ecosystem.

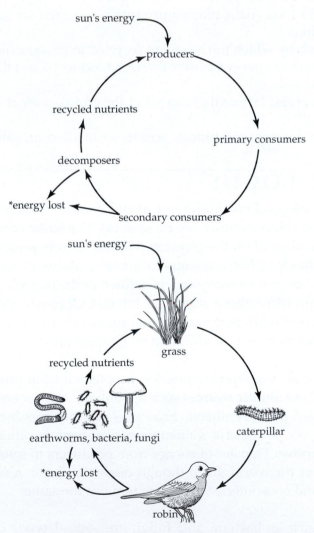

*Indicates that energy is lost at each step of the cycle.

FIGURE 21-1. Transfer of Energy and Nutrients through Ecosystems

ACTIVITY 1 FOOD CHAINS

The food molecules that algae, green plants, and other producers make during photosynthesis form the base of a food chain that includes all living organisms (plants, animals, protists, fungi, and bacteria).

These food chains illustrate an important biological principle: all organisms are linked together for survival and are therefore interdependent. Humans can't separate themselves from these interactions. The more we understand about how living organisms interact, the better prepared we'll be to ensure that these cycles continue.

To demonstrate the transfer of energy in a simple terrestrial food chain, consider **Figure 21-2.** The producer in this food chain is a plant, corn, which performs photosynthesis. Each consumer level is identified in sequence. For example, the **worm** is referred to as a **primary consumer,** the **blackbird** as a **secondary consumer,** and the fox as a **tertiary consumer.** Dead organisms and waste materials from any level of the food chain become an energy source for **decomposers.**

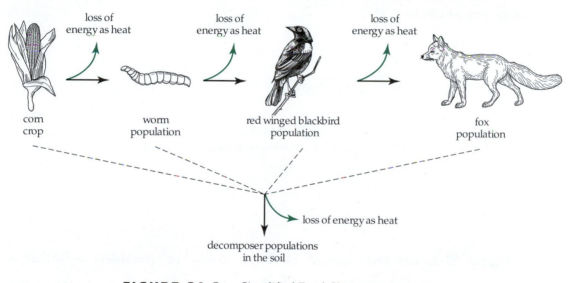

FIGURE 21-2. Simplified Food Chain in an Ecosystem

Consumers are also classified according to their food choices. For example, consumers that eat only producers are called either primary consumers or **herbivores.** Consumers that eat only other animals are called **carnivores** (and may feed at different levels of the food chain). Many consumers have a varied diet that includes both plant and animal foods. These are referred to as **omnivores.**

However, energy transfer through food chains isn't an efficient process. Your body isn't able to utilize all the calories present in food plants or animals. Conversion of useful energy to **heat** is one way that energy can be lost in food chains. The **loss of heat energy** is shown in **Figure 21-2. Energy loss occurs at every level of the food chain.** In general, we consider that only **10%** of the available energy in one level of the food chain (called a **trophic level**) is available to feed consumers in the subsequent level.

1. On **Figure 21-2,** label each organism with its appropriate level in the food chain: **producer, primary consumer, secondary consumer,** and **tertiary consumer.**

2. If the red winged blackbird ate the corn directly (instead of eating worms), what level consumer would it be? _____

3. In **Figure 21-2,** which organism has the **least** available energy to support its population? _____

4. Which organism in **Figure 21-2** takes energy directly from sunlight? _____

5. **Not** including the decomposers, how many trophic levels are indicated in **Figure 21-2?** _____

6. If the red winged blackbird dies from natural causes, what will happen to the **energy** present in its body?

7. What will happen to the mineral nutrients released to the soil by decomposers?

8. In **Figure 21-2,** the red winged blackbird could be classified as either a _____ or a _____ according to its position in the food chain.

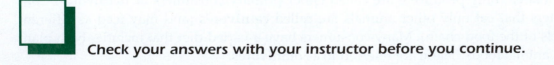

Check your answers with your instructor before you continue.

ACTIVITY 2   LOSS OF ENERGY IN FOOD CHAINS

1. Work in groups. From the supply area get **a liter flask, a bottle of food coloring, one graduated cylinder, one graduated 10-ml pipette with manual dispenser, one graduated 1-ml pipette with manual dispenser, and four small cups.**

2. Fill the flask with **one liter** of tap water. Add enough food coloring until the liquid is **brightly colored** and gently shake to mix.

    The water in the flask **represents solar energy** that reaches the earth.

3. From the flask of colored water, measure **100 ml of fluid into the graduated cylinder** and place it in the **first cup.**

    What **percentage** of the total volume of liquid was removed? _____ %

    (Circle one answer.) Which **level of the food chain in Figure 21-2** does **cup #1** represent? **producer / primary consumer / secondary consumer / tertiary consumer**

Note:

Your instructor will demonstrate the use of the graduated pipettes that you'll need to complete steps 4 through 6.

---

4. **From cup #1,** remove **10 ml of fluid with the 10-ml pipette** and place it in **cup #2.**

    What **percentage** of the total volume of liquid was removed? _____ %

    (Circle one answer.) Which **level of the food chain in Figure 21-2** does **cup #2** represent? **producer / primary consumer / secondary consumer / tertiary consumer**

5. **From cup #2,** remove **1 ml of fluid with the 1-ml pipette** and place it in **cup #3.**

    What **percentage** of the total volume of liquid was removed? _____ %

    (Circle one answer.) Which **level of the food chain in Figure 21-2** does **cup #3** represent? **producer / primary consumer / secondary consumer / tertiary consumer**

6. **From cup #3,** remove **0.1 ml (one-tenth ml) of fluid with the 1-ml pipette** and place it in **cup #4.**

    What **percentage** of the total volume of liquid was removed? _____ %

    (Circle one answer.) Which **level of the food chain in Figure 21-2** does **cup #4** represent? **producer / primary consumer / secondary consumer / tertiary consumer**

7. Consider the original flask, which contained one liter of fluid.

    How much fluid remains in the flask? _____ ml

    What **percentage** of the original fluid remains in the flask? _____ %

8. Place the flask and the four cups in a straight line on the table. Consider the volume of liquid in the flask and the cups.

    If the volume in each container represents energy, **summarize the pattern of energy flow** through a **food chain.**

9. Based on your answer to **question #8,** would you expect to see a food chain with 10 levels in a normal ecosystem? **Explain** your answer.

10. What does the colored water remaining in the flask represent in your ecosystem?

Check your answers with your instructor before you continue.

> ### Note:
>
> In this demonstration, the corn crop was able to absorb 10% of the incoming solar energy. In real life, the percentage is much lower than this (closer to 1%). However, the amount of solar energy that reaches planet earth is many, many times greater.

ACTIVITY 3

FOOD CHAINS INTERACT TO FORM FOOD WEBS

The simplified food chain that you studied in Activities 1 and 2 isn't a true representation of energy transfer in natural ecosystems. This should make sense, as you know that **most animals eat more than one type of food,** depending on their age, the season of the year, and so on. These complex relationships are represented by a **food web.**

Figure 21-3 contains a **simplified version** of a food web that might be found outside your city. As in the food chain in **Figure 21-2,** the arrows on the diagram indicate the direction of energy transfer (for example, when the fox feeds on the mouse, the fox receives energy and nutrients from its food). **Decomposers exist in food webs,** but they often aren't shown, because the arrows are difficult to position in the diagram.

Based on the basic principles of food chains discussed in the previous activities, answer the following questions about the food web in **Figure 21-3.**

1. How many types of producers are represented in this food web? _____

2. How many types of consumers are represented in this food web? _____

3. Which organisms are primary consumers?

4. Which organisms are secondary consumers?

5. Which organisms are tertiary consumers?

6. Which organism(s) function as **both** primary and secondary consumers?

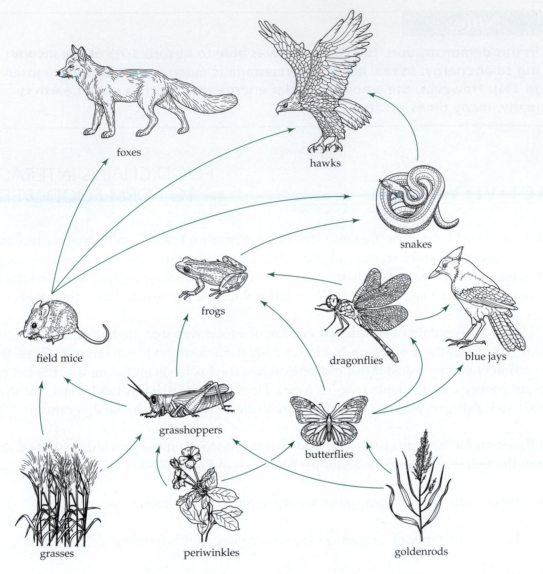

FIGURE 21-3. Simplified Terrestrial Food Web

7. If decomposers were included in this food web, which organisms in the food web
    would have arrows leading to decomposers?

Challenge Questions!

8. Imagine that the grass population in this ecosystem contains 1,000,000 calories of energy. In the food chain involving the grasses, field mice, and hawks, how many of the calories found in the grass population would be available to feed the hawk population? **Show your work.**

9. How many different food chains begin with the periwinkles? _____
    List them below:

ACTIVITY 4   ECOSYSTEM JENGA®

Adapted from Whale Jenga by Cabrllo Marine Aquarium (cabrillomarineaquarium.org)

Food webs exist not only in terrestrial ecosystems, but also in aquatic ecosystems. When considering a large marine ecosystem such as the ocean, it's difficult to envision what's happening below the surface because many of the organisms that are involved in these food webs aren't easily visible. Also, the world's oceans are large and interconnected, resulting in vast bodies of water. Because the oceans are so large, is it possible that human actions, even small actions, can affect these large, complicated ecosystems?

   Changes in biotic and abiotic factors can impact different components of the food web and these changes alter a marine ecosystem. To see an example of how this works, play the following game that simulates the **marsh and marine ecosystems off the coast of Louisiana.**

1. Work in groups. From the supply area, get **one Jenga® game with colored blocks.** Remove the **game card pages from Appendix 1.**

2. Set up the Jenga pyramid as follows:

    ■ Lay the plastic loading tray on the table—the solid end of the tray will form the bottom of your tower after it's built.
    ■ Starting at the solid end of the tray, place three blocks in each layer at right angles to the previous layer.
    ■ Place all the **green blocks** in the bottom layers.
    ■ When all the green blocks have been stacked, continue the process with the **red blocks** until they're all stacked as well.
    ■ Stack all the **blue blocks** next and finish with the **yellow blocks.**
    ■ There should be no blocks left over when the tower is complete.

3. Using the loading tray for support, set the tower upright. Carefully remove the tray from the tower.

4. The organisms in your simulated marine ecosystem are listed in **Table 21-1**.

> ### Note:
>
> **For this example, although it's understood that decomposers are very important in any food web, they aren't being shown in our sample ecosystem.**

TABLE 21-1

TROPHIC LEVELS OF A TYPICAL LOUISIANA COASTAL ECOSYSTEM

TROPHIC LEVEL	BLOCK COLOR	REPRESENTATIVE ORGANISMS
Producers	Green	phytoplankton (tiny photosynthetic producers including algae and photosynthetic bacteria)
		sea grasses
		marsh grasses
		larger algae (seaweeds)
Primary consumers	Red	small fishes (sardines, menhaden)
		oysters and mussels
		zooplankton (small floating animals including krill, copepods, shrimp, and crab larvae)
		ducks
		fiddler crabs (small shore crabs—the male has one big claw)
Secondary consumers	Blue	larger fishes (tuna, red snapper)
		squid
		shrimp and blue crabs
		pelicans
		whale sharks
		people
Tertiary consumers	Yellow	blacktip sharks
		dolphins
		sperm whales
		people

5.   Cut the game cards apart and arrange the cards in a stack with the **instructions face down.** Place the cards in numerical order so that **card #1 is on top** of the stack.

6.   The first player picks card #1, reads it aloud, and follows the instructions written on the card.

   Jenga rule: You can gently touch other blocks to find a loose one in the stack, but you can't hold the rest of the stack together while removing your block.

7.   Once a card has been used, it won't be used again, so set the used cards aside. Set the removed blocks to the side (far enough away from the tower so that they won't be included in the pile when the tower collapses).

   Continue to take turns until the tower collapses or all the game cards have been used.

✔ Comprehension Check

1.   In regard to the information on **game card #7:** Based on your knowledge of food chains, why did this action result in the addition of a green block, but the removal of a blue block?

2.   In regard to the information on **game card #10:** Why did this accident result not only in the removal of two green blocks but also the removal of a yellow block?

3. In regard to the information on **game card #13:** Why did this population shift result in the addition of a blue block to the tower?

4. In regard to the information on **game card #18:** Why did this action result in the addition of a green block?

   Was the addition of the green block due to a biotic or an abiotic factor? Explain your answer.

5. Assume that your leftover yellow blocks (the tertiary consumers) represent only people. Why was there less effect on that trophic level than on the lower levels of the food chain?

6. **Challenge Question!** You removed a large number of blocks and the tower remained intact. But, the removal of just one additional block caused the collapse of the entire tower. What ecosystem principle does this demonstrate?

 Check your answers with your instructor before you continue.

SELF TEST

1. List three examples of producers that were not mentioned in this laboratory exercise.

2. Draw a food chain for a vacant city lot that has weeds, rats, cats, and crickets. Label each member of the food chain according to whether it is a producer, a primary consumer, and so on.

3. If one of the cats dies of natural causes, what will happen to its remains? What will happen to the mineral nutrients contained in its body? Explain your answers.

4. Complete the crossword puzzle in **Figure 21-4** (page 393) to review the basics of ecosystem structure and function.

ACROSS
  4. Microscopic organisms found at the beginning of aquatic food chains
  6. In a typical ecosystem, there are ____ herbivores than carnivores
  7. Chemical and physical factors present in an ecosystem
  9. Organisms that eat only producers are called ____ consumers
 11. Percentage of energy decrease at each link of a food chain
 13. Organism that makes food energy from nonliving sources
 14. In a food chain, each trophic level passes on ____ energy than it receives
 16. Based on their diet, most people would be classified as ____
 17. The number of trophic levels there would be in a food chain that received
       100,000 kilocalories of energy from the sun and ended with a level that had
       100 kilocalories

DOWN
  1. Process of changing solar energy into food energy
  2. The number of organisms in a food web that could become a food source for
       decomposers
  3. Organism most responsible for nutrient recycling in soil or water
  5. In the following food chain (phytoplankton → sardine → tuna → dolphin),
       the tuna would be classified as a ____ consumer.
  8. Percentage of energy available between links in a food chain
 10. Term that describes a biological community and the physical factors in that area
 12. Source of energy for photosynthesis
 15. Interconnected network of food chains

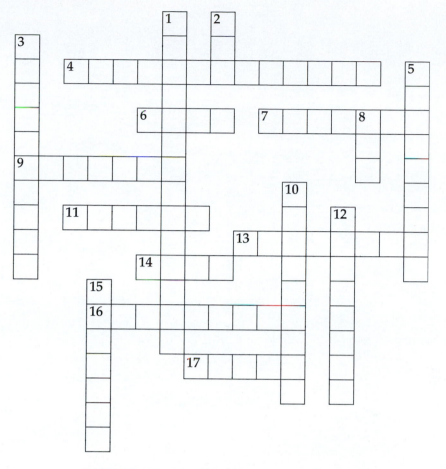

FIGURE 21-4. Ecosystem Crossword Puzzle

Population Ecology

Objectives

After completing this exercise, you should be able to:

- estimate the size of a population using various methods of sampling
- use ecological quadrats as a tool for predicting population size
- calculate the percent error of your estimate
- draw conclusions that are supported by experimental data
- apply your knowledge of the scientific method to real-life situations

CONTENT FOCUS

In studying a population in an ecosystem, it's essential to know the size of the population in order to analyze resource availability. Examples of resources needed by individuals of an animal population include living space, food, water, and shelter. Plants need access to sunlight, mineral nutrients, suitable soil, and growing space.

Population **density,** the number of individuals in a given area, determines the availability of resources to the individuals of the population. Individuals may be distributed randomly, uniformly, or in clumps. This makes sense when you consider that in nature, resources aren't always distributed evenly in the environment.

Ecologists call the total count of all the individuals in a population a **census**. However, scientists can't possibly count every organism in a population. Fortunately, other methods are accurate enough for most research purposes. One way to estimate the size of a population is to collect data by taking random samples. Sampling tracks changes in population patterns over time. Plant researchers often use sample plots called **quadrats**. Animal researchers may use mark and recapture techniques or other counting methods.

In this exercise, you'll try several different methods of estimating population size and distribution and then compare the data collected by random sampling with data obtained by an actual count.

Thinking About Biology ■ **Exercise 22**

ACTIVITY 1
INTRODUCTION TO SAMPLING—
AN ECOLOGICAL TOOL

Estimates of population size are basic to understanding the interactions of populations in an ecosystem. It's seldom practical, however, to count every individual in a population. Even if possible, direct counts may be too time consuming or expensive. Sampling is a way to make a quick estimate of population size. To give you an idea of how an ecologist would sample a wild population, we've taken a photograph of a flock of birds.

1. Work in groups. Consider the population in **Figure 22-1.** It would be hard to count the exact number of birds in this flock, especially if the flock was moving, so a quick estimate is our best option.

2. Look at the population in **Figure 22-1.** If you stretch your imagination, you'll see that the area covered by birds can be roughly divided into six squares.

   DON'T draw grid lines on the photograph—you're simulating what you would be able to do if you were actually outside counting the birds!

3. To make a population estimate, count the number of birds in one square and multiply the total by six.

   _____ × 6 = _____ (total population)

4. How close did you come to the actual population? Count every bird in the photograph to find out.

   Exact count of total population = _____

   Check one of the following to rate your estimating accuracy:

   Within **5 birds** of the actual total   Accuracy **100%**

   Within **25 birds** of the actual total   Accuracy **90%**

   Within **50 birds** of the actual total   Accuracy **80%**

   Within **70 birds** of the actual total   Accuracy **70%**

   Within **100 birds** of the actual total   Accuracy **60%**

✓ Comprehension Check

1. How does each of the following factors affect the accuracy of population estimates? **Explain** each answer.

   a.   Practicing making estimates:

   b.   Making sure your "counting" square contains a representative number of animals:

   c.   Averaging the results of two squares or three "counting" squares:

2. What are the advantages of estimating population size using the "imaginary counting square" method? What are the disadvantages?

Check your answers with your instructor before you continue.

FIGURE 22-1. Birds on the Wing

ACTIVITY 2

SAMPLING QUADRATS TO ESTIMATE POPULATION

You're a student in an ecology class that has taken a field trip to Yellowstone National Park. Each grid square in **Figures 22-3 and 22-4** represents a study area within the park. Each study area is divided into 100 **quadrats** for vegetation sampling (see **Figure 22-2** for an example of how a quadrat is established).

FIGURE 22-2.   Example of a Quadrat

You'll be engaged in a habitat assessment of the meadow community in the study area. Habitat assessment is a common tool used to obtain an overall evaluation of a habitat at relatively low cost. With baseline data obtained by habitat assessment, we can predict the environmental impact of activities proposed for this location.

We're beginning with an assessment of plant populations. Since plants are fixed in position, we can establish measured grids to facilitate estimates of population size and distribution.

Building on the random sampling method you practiced in Activity 1, you'll try to make an accurate estimate of the population of gold flowers (*Hymenoxys acaulis*).

1.   Work in groups. Get **one set of ten alphabetic chips and one set of ten numeric chips.** Each set of chips will be in a small container.

2. To begin sampling, randomly remove **one chip from each container.** Record the number and letter you drew in **Table 22-1** in the column titled **Quadrats Sampled in Figure 22-3,** for example: **F8.**

   Return the chips to the container before you draw another chip combination.

   Continue drawing chips from the two containers and recording the numbers until you've identified **10 quadrats to sample.** If you draw a duplicate letter/number combination, return both chips to their containers and draw again.

3. To sample the first quadrat on your list, locate the quadrat in your grid square. **Count all the plants in that square** (represented by dots on the grid) and **record the total** next to the appropriate quadrat number in **Table 22-1.**

   Continue with this procedure until you've sampled and recorded the data for all 10 quadrats.

Note:

What should you do in case a dot is on the line between two quadrats? If at least half of the dot is in the quadrat you are sampling, count it as part of your total for that quadrat.

If you count a dot that's on a line, mark through that dot so that you don't mistakenly count it again as part of the population in an adjoining quadrat.

---

4. **Repeat Steps 2 and 3** above to sample the population in **Figure 22-4.**

   Return to **Table 22-1** and **record the quadrat numbers** you'll sample in the column titled **Quadrats to Sample in Figure 22-4. Record your population results** in the **Plants Counted** column for **Figure 22-4.**

5. **Record the total number of plants counted in all 10 quadrats** and record the data in **Table 22-1.**

   To estimate the total population size in your grid square, **multiply the total number of plants counted by 10. Record your population estimate** in **Table 22-1.**

6. Return to the grid square and **count every dot on the grid** (an actual count of the population present in your grid square).

   Record the actual population count in **Table 22-1.**

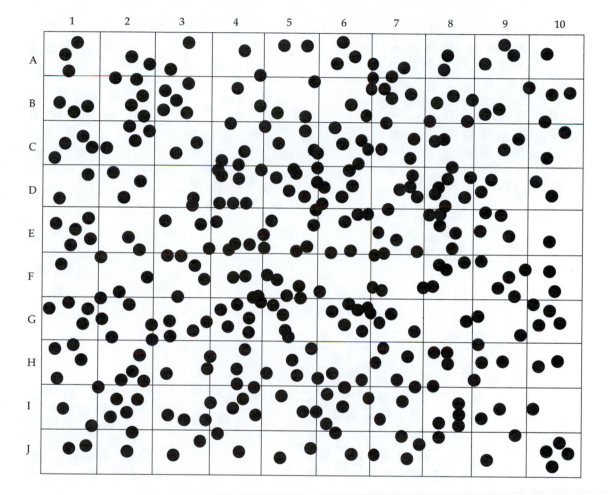

FIGURE 22-3. Study Area 1

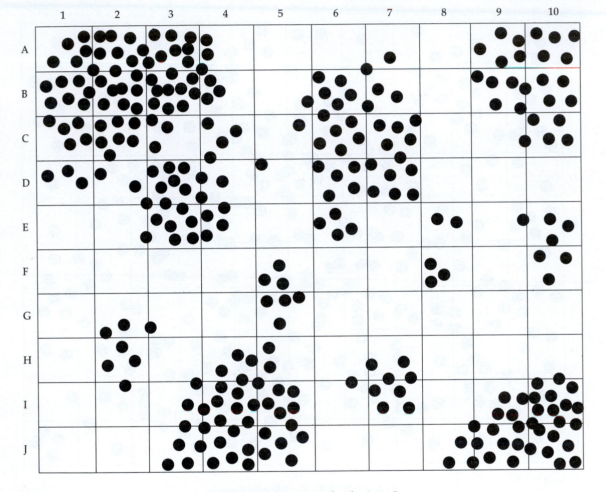

FIGURE 22-4. Study Area 2

7. Use the following formula to **calculate the percent of error** between your population estimate and the actual population:

$$\text{percent error} = \frac{\text{actual population} - \text{estimated population}}{\text{actual population}} \times 100$$

Record your **percent error** in **Table 22-1**.

TABLE 22-1			
VEGETATION SAMPLING DATA			
QUADRATS SAMPLED IN FIGURE 22-3	PLANTS COUNTED IN FIGURE 22-3	QUADRATS SAMPLED IN FIGURE 22-4	PLANTS COUNTED IN FIGURE 22-4
	Total Plants Counted:		Total Plants Counted:
	Population Estimate from Sample:		Population Estimate from Sample:
	Actual Population Count:		Actual Population Count:
	Percent Error of Estimate:		**Percent Error of Estimate:**

☑ Comprehension Check

1. Was there a difference in the percent error of your population estimate between **Figures 22-3 and 22-4?** _____

   If so, how might the following factors have contributed to the difference?

   a.   Distribution of plants in the quadrats (evenly distributed vs. patches of plants):

   b.   Number of quadrats sampled:

2. Why did we use the "chip" method to select the quadrats to be sampled instead of simply selecting 10 "likely looking" quadrats?

3. List and explain several environmental factors that might contribute to the patchy distribution of the plants in **Figure 22-4.**

4. List **three situations** in which **random sampling of human populations** is used to collect information.

5. **Challenge Question!** While passing by your local elementary school, you notice that almost all the parents waiting to pick up their children are driving SUVs. You wonder if the same distribution of vehicles would be observed at other elementary schools in the United States.

   Design an experiment that uses sampling techniques to answer this question. **Explain** your method and the reasoning behind your experimental design in detail.

Check your answers with your instructor before you continue.

ACTIVITY 3   THE MARK AND RECAPTURE TECHNIQUE FOR ESTIMATING POPULATION SIZE

If you're trying to sample a population of animals that can move from place to place, the quadrat system won't be very effective. To sample this type of population, individuals are captured in a study area, marked for identification, and released. Later, individuals will again be captured in the same study area. Some of these newly captured individuals will be marked; others will not. On the basis of the ratio between marked and unmarked individuals captured, it is possible to make a mathematical estimate of the population size.

The following activity is a simulation of the mark and recapture technique demonstrated by sampling a captive population of brine shrimp (see **Figure 22-5**).

FIGURE 22-5. Brine shrimp (*Artemia salina*)

1. Work in groups. Get **a container of brine shrimp in clear saline solution (labeled "brine shrimp population"). You'll also need a small beaker of brine shrimp that have been colored with methylene blue dye, several disposable pipettes, and two empty beakers.**

   Label the first beaker #1 and the second beaker #2.

2. **Capture Phase**

   Capture **30 brine shrimp** using the following method.

   To "capture" the brine shrimp, fill a pipette in a **random location** in the "population" container.

   When you capture a sample of brine shrimp, you'll place the captured individuals in **beaker #1.**

3. **Marking Phase**

   Since it would be almost impossible to mark an animal as small as a brine shrimp during one lab period, we'll use an alternate method of releasing marked individuals into the population.

Replace each brine shrimp captured and removed from the population with a blue brine shrimp (you'll be releasing a total of 30 blue individuals into the original brine shrimp population).

Wait for five minutes to allow the marked individuals to become randomly dispersed in the population.

4. **Recapture Phase**

Your recapture sampling of the population **MUST BE RANDOM.**

WITHOUT LOOKING, insert the pipette into the population container and draw up a sample.

Count the number of marked and unmarked brine shrimp in the pipette.

Note:

If counting the shrimp in the pipette proves difficult, release the contents of the pipette into beaker #2 and then count the shrimp. By placing the beaker on a white sheet of paper, the brine shrimp will be easier to observe and count.

---

5. **Record the number of marked and unmarked individuals captured in the first pipette in Table 22-2.**

6. After you've recorded your shrimp count, transfer the counted shrimp to **beaker #1.**

    DON'T REPLACE THE SAMPLED SHRIMP IN THE "POPULATION" CONTAINER.

    Continue taking random recapture samples and recording your data until you've recaptured **10 marked individuals.**

	TABLE 22-2		
	RESULTS OF MARK AND RECAPTURE EXPERIMENT		
PIPETTE SAMPLE TAKEN	NUMBER OF UNMARKED SHRIMP CAPTURED	NUMBER OF MARKED SHRIMP RECAPTURED
1.		
	Total:	Total:

7. Use the following formula to calculate your population estimate:

   N = total population size

   C = total number of shrimp captured (marked and unmarked)

   M = total number of shrimp marked

   R = number of marked shrimp recaptured

   $$N = \frac{(C)(M)}{R}$$

   On the basis of your calculations, what is your shrimp population estimate?
   _____ **individuals**

8. How accurate was your population estimate? Several factors affect the degree of confidence in your results. The accuracy of your population estimate will rise, for example, if you sample the population more often. If you were actually participating in a scientific population study, the degree of accuracy would be calculated based on a principle known as the **95% confidence interval.** The confidence interval calculations produce a range of numbers (high and low) between which the actual population figure should lie (with 95% confidence that this is correct).

✓ Comprehension Check

1. **(Circle one answer.)** If the marked individuals were more visible to predators than the unmarked individuals, my population estimate would be **too high / too low.**

   Explain your answer.

2. **(Circle one answer.)** If marked animals migrated out of my study area, my population estimate would be **too high / too low.**

   Explain your answer.

3. Fishery managers have used population estimates to determine the maximum sustainable yield (the degree of harvesting that doesn't harm the populations' long-term survival). What would be the benefits and the shortcomings of using the **mark and recapture technique** to estimate the number of tuna that commercial fishermen are allowed to capture each year?

Check your answers with your instructor before you continue.

SELF TEST

Fill in the blank with the most appropriate choice. Answers can be used **only once**.

a. quadrat
b. mark and recapture
c. too high
d. too low
e. 95% confidence interval
f. sampling

1. _____   If a sampled quadrat includes a clump, your population estimate might be _____.

2. _____   A method of estimating population size based on trapping and tagging individuals in a small sample.

3. _____   Measurement of the reliability of data analysis.

4. _____   If you're making a quick estimate of population size, but the animals are moving really quickly, your population estimate may be _____.

5. At a forested site, a lumber company sampled the number of 12-inch diameter pine trees in preparation for logging. Later, in a direct count, it was discovered that the company had seriously underestimated the number of suitable trees. What factors may have contributed to the lack of estimating accuracy? **Explain** your answer.

6. Several state and federal agencies support programs to tag and release commercially important fishes. For example, sport fishermen tag and release the large fish they catch. If someone catches a tagged fish at a later date, they are asked to report the geographic location, the size of the fish, time of the year, and other information. How can this information be used to benefit the tagged species?

Each of the following sampling techniques has been used to determine population density. For each method, explain how this approach would let you determine the population size.

7. Number of tuna caught during 100 hours of fishing with nets.

8. Number of frog calls heard every 10 minutes on a spring evening over a two-hour period.

9. Number of roaches caught in traps in a one-week period in an apartment complex.

Game Cards for Exercise 21

1 Changes in ocean currents disperse phytoplankton. Remove two green blocks.	**5** Nutrient runoff from overuse of fertilizers causes a toxic algal bloom. Remove one red block and one blue block.
2 Runoff from huge rainstorm reaches the marsh and ocean, decreasing the salinity of the water. Remove one blue block and one red block.	**6** Smoke from burning oil wells blocks sunlight over ocean. Remove two green blocks.
3 Whale sharks spend longer than usual feeding in the area. Remove one red block.	**7** Overharvesting of oysters and mussels decreases the populations. Insert one green block back into the matching color level and remove one blue block.
4 Oyster repopulation program is highly successful. Remove one green block.	**8** College students conduct a large beach cleanup campaign. Insert one green block and one red block back into the matching color levels.

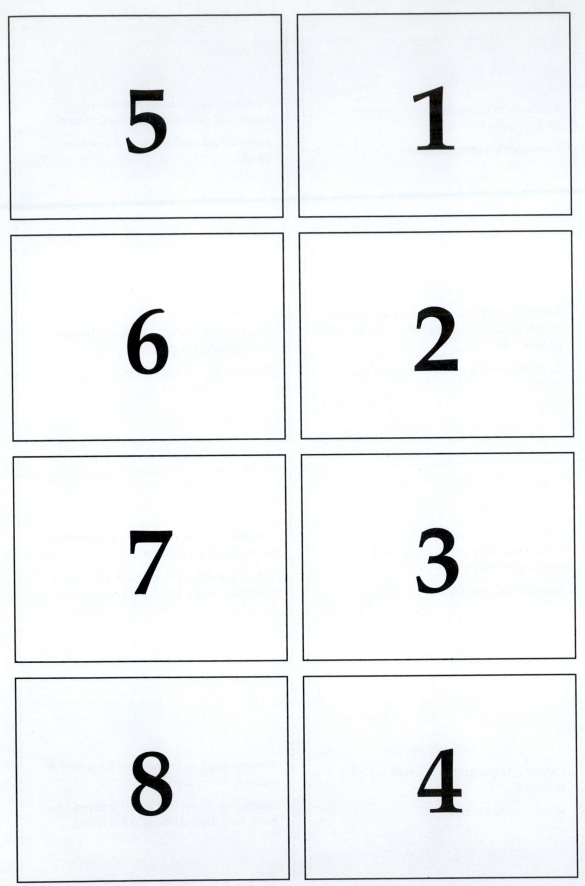

	9
Large dolphin school passes through the area. Remove one blue block.	

	13
Hurricane Katrina causes a population shift away from Louisiana to other states. Remove one yellow block and insert one blue block back into the matching color level.	

	10
Oil spill offshore is carried into the marshes by prevailing currents. Remove two green blocks and one blue block.	

	14
Blue crab population along the mid-Atlantic states decreases significantly. More crabs are shipped north from Louisiana. Remove one blue block.	

	11
After the oil spill, sick and dying dolphins wash up on shore. Remove one yellow block.	

	15
There's a population boom in sardine and menhaden fry (newly hatched juveniles). Remove two green blocks.	

	12
Unusually cold winter in the northern states causes ducks to arrive on the Louisiana coast much earlier than usual. Remove two green blocks.	

	16
Use of DDT pesticide prevents pelican eggs from hatching. Remove one blue block.	

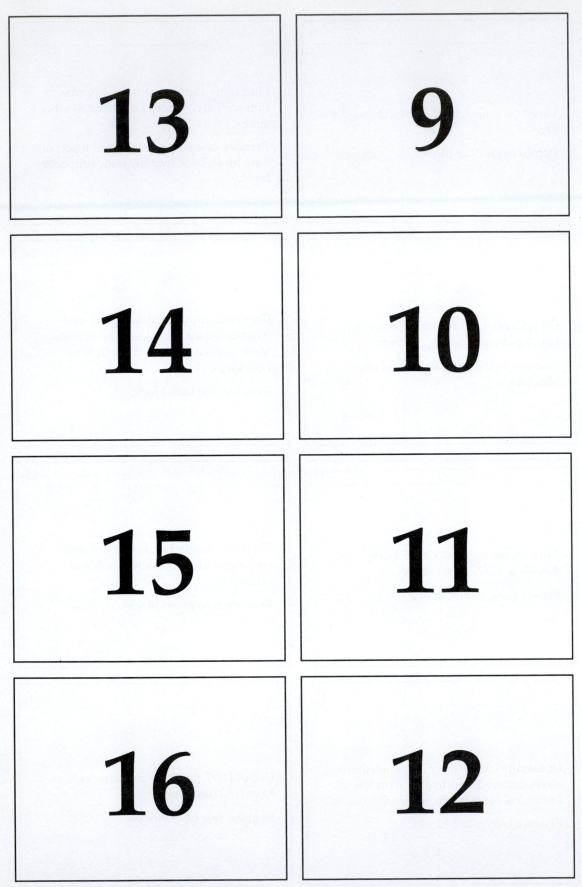

17

Huge flocks of migratory birds pass through the marsh ecosystem on the way to their breeding grounds.

Remove one red block.

20

Oil removal efforts in the marshes take longer than expected.

Remove one green block.

18

Lots of clear sunny days. More sunlight reaches the ocean surface.

Insert one green block back into matching color level.

21

Thinning of the ozone layer lets more damaging ultraviolet radiation reach the ocean surface.

Remove one green block.

19

Fishermen exceed fishing regulations and bring in a huge tuna harvest.

Remove one blue block and insert one red block back into the matching color level.

Self Test Answers

Exercise 1—Introduction to the Scientific Method

1. C—experimental

2. B—control

3. E—verify

4. G—Y axis

5. J—average

6. H—data point

7. A—variable

8. a. The points are too closely spaced (they should be spread out as much as possible along the Y-axis) and crowded into the corner (they should be evenly spaced along the X-axis).

    b. No numbers or information about scale on the X-axis

    c. No graph title

9. a. Titles missing for the X- and Y-axes

    b. Numbers on the Y-axis do not have equal intervals; the scale changes from intervals of 50 (50–100) to intervals of 500 (100–500)

    c. No numbers or information about the scale on the X-axis

    d. Plotted points extend above the number scale on the Y-axis

10. Decreased

11. About 13%

12. About 88%

13. 2001

14. Your graph should look similar to the following example.

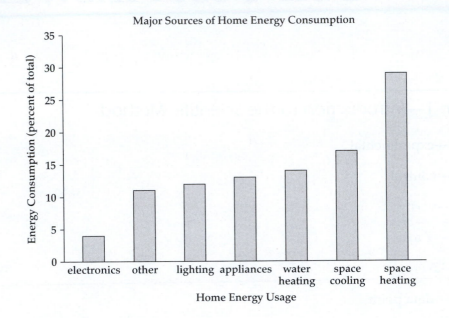

Major Sources of Home Energy Consumption

Exercise 2—Windows to a Microscopic World

1.

Ocular lens (eyepiece)	*Allows you to view the specimen; magnifies image 10×; contains the pointer*
Stage	*Holds the specimen; has a mechanism to move the slide around*
Condenser lens	Focuses light on the specimen
Nosepiece	*Revolves to change objective*
Scanning lens	Objective lens used first to locate a specimen
Iris diaphragm	Regulates the amount of light that passes through the specimen
Fine focus knob	*Moves objective lenses in small increments for small adjustments in image clarity*
Scanning lens (4X)	Objective lens with the lowest magnifying power

High power (40X)	Objective lens with the highest magnifying power
Coarse focus knob	*Moves objective lenses rapidly in large increments for initial adjustment in image clarity*

2.   Total magnification: $15 \times 20 = 300$

3.

- In the compound microscope, the light source is beneath the specimen. Therefore, only thin specimens can be viewed.

- The dissecting microscope can provide light from many different directions so that large, thick objects can be viewed.

- In the compound microscope, the image is viewed upside down and backward. This is not true for a dissecting microscope.

- The dissecting microscope has a zoom lens that can gradually increase magnification. The lenses of the compound microscope have fixed magnification levels.

- The compound microscope has much greater magnification ability than a dissecting microscope.

4.   Compound—cells from the lining of your stomach

Dissecting—a seashell you found on the beach

Dissecting—a cockroach you found in the kitchen

Compound—mold from your shower curtain

5.   The daphnia moved away from you and to your left (because images in the compound microscope are inverted and reversed).

6.   a.   Go back to the lens and center the specimen in the middle of the field of view or check to make sure the objective is clicked into position.

        b.   Adjust the iris diaphragm to let in more light.

        c.   Check to make sure the objective lens is clicked into position, or check to see if the objective and ocular lenses are clean.

        d.   Use lens paper to clean the slide, the objective lenses, and the ocular lens.

        e.   These are air bubbles. Lift the cover slip and lower it gently at a 45° angle to expel the air.

Exercise 3—Functions and Properties of Cells

1.   Central (sap) vacuole

2.   Lipids

3.   Nucleus

4.   Protein synthesis

5.   Chloroplasts and central (sap) vacuole

6.   Ribosomes, endoplasmic reticulum, Golgi apparatus

7.   Chloroplasts

8.   Mitochondria

9.   Cilia

10.   Chloroplasts, central (sap) vacuole, cell wall

11.   As the equation for photosynthesis shows, water is a critical component for the process. Central vacuoles provide a storage reservoir for water in case of drought or other types of low-water conditions.

12.   Since aerobic cellular respiration occurs in the cell's mitochondria, this would be the primary organelle affected by the cyanide gas.

Exercise 4—Movement of Molecules Across Cell Membranes

1.   The arrow should show water moving from the bag into the beaker.

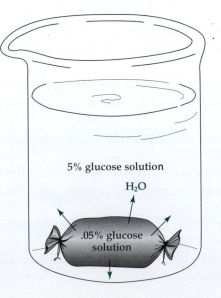

5% glucose solution

H_2O

.05% glucose solution

2. Dialysis is the separation of different-sized molecules through a selectively permeable membrane. As diffusion takes place across the membrane, the composition of the blood changes. Waste products such as phosphate ions, urea, and potassium ions cross the membrane into the dialysis fluid. Blood cells, proteins, and other large molecules cannot cross the membrane and are retained in the blood.

3. Distilled water has no dissolved solutes. Osmosis refers to the movement of water molecules from an area of higher concentration to an area of lower concentration. In the described situation, there is a lower concentration of water molecules inside the cell (in the cytoplasm) and a higher concentration of water outside the cell (the distilled water). Through the process of osmosis, the distilled water will cross the cell membrane and enter the blood cells, causing them to swell and rupture the cell membrane.

4. The sugar cube will dissolve faster in the hot tea. During diffusion, molecules move from an area of higher concentration to an area of lower concentration. Because the water is hot, molecular movement increases. Collisions between molecules become more frequent. As molecules bump into each other, they diffuse outward, spreading from the area of higher sugar concentration (the sugar cube) to an area of lower sugar concentration (the cup).

5. The smoke molecules are moving by diffusion over the restaurant barrier from an area of higher smoke concentration (the smoking section) to an area of lower smoke concentration (the nonsmoking area).

Exercise 5—Investigating Cellular Respiration

1. a. The cloudy result indicates that the gas obtained from the body cavity contained CO_2.

    b. CO_2 is a by-product of cellular respiration. Cell respiration occurs only in living organisms. Because the animal is dead, the carbon dioxide that was produced must be coming from the cellular respiration of other organisms (decomposers present on the carcass).

2. A—aerobic respiration. Aerobic respiration produces about 36 ATPs per glucose molecule, whereas anaerobic forms of cellular respiration produce only 2 ATPs per glucose molecule.

3. Bromothymol blue returned from yellow to its original blue color because CO_2 was no longer present. Because plants absorb CO_2 during photosynthesis, and the plant was exposed to sunlight during the several hours in question, this is evidence that photosynthesis occurred.

4. B—only under anaerobic conditions. In the absence of sufficient oxygen, yeast will shift their method of cellular respiration to an anaerobic method—alcohol fermentation. In this way, they can continue to obtain ATP to support cellular activities.

5.   C—under both aerobic and anaerobic conditions. Breakdown of glucose for cell respiration begins with glycolysis, an anaerobic process. Cellular respiration will continue as shown in Figure 6-3, but the pathway used is determined by the presence or absence of oxygen.

6.   B—only under anaerobic conditions. During strenuous exercise, when sufficient oxygen is not available to muscle cells, cellular respiration shifts to an anaerobic method—lactic acid fermentation. In this way, muscle cells can continue to obtain ATP to support cellular activities.

7.   A—aerobic conditions only. Under ideal conditions, cells can maximize their ATP production.

8.   D—Even though anaerobic respiration produces small amounts of ATP, the amount is not sufficient to meet the energy needs of an animal for long periods of time.

Exercise 6—Photosynthesis

1.   Examples: algae, phytoplankton, pine tree, rose bush, corn, rice, beans, photosynthetic bacteria (any organism that contains chlorophyll is acceptable).

2.   Some plants store starch in their roots. Examples include potatoes, carrots, radishes, turnips, and beets.

3.   No, the number of stomata should be greatly reduced. This would help reduce loss of water through evaporation.

4.   If you increased the amount of light available to the elodea plants, the rate of photosynthesis would increase. Another option would be to increase the CO_2 concentration in the water, by blowing through the straw for a longer period of time. These two options would increase photosynthesis because the equation shows that both light and carbon dioxide are necessary for the process to occur.

5.   No. Although enough light would be available for photosynthesis, the CO_2 in the test tube would gradually be depleted, slowing the rate of photosynthesis. Additional CO_2 from the atmosphere can't diffuse into the test tube because of the stopper closing the tube.

6.   a.   The stomata should be open because the moist, cool temperature minimizes water loss and the sun is out for photosynthesis.

                b.   The stomata should be closed to reduce water loss.

7.   On the basis of the methods used in this experiment, the conclusions are not valid. When the refrigerator door is closed, the light is out. Because two variables are involved (temperature and light), it is impossible to tell which caused the observed effects.

Exercise 7—Organic Molecules and Nutrition

1.   Your morning orange juice doesn't contain protein.

2.   Because potatoes contain large amounts of starch, the iodine would probably turn black (a positive starch test). Because animal tissues do not contain starch, the iodine test should be negative (no color change).

3.   No, the experimental conditions weren't valid. Benedict's test identifies the presence of simple sugars, but the student added glucose solution to the test tube. Glucose is a simple sugar, causing a false positive in the results. In reality, the presence of sugar in the potato was never tested.

4.   Answers will vary, but in most cases, a shift from eating fatty foods to consuming more fruits, vegetables, and complex carbohydrates would have positive effects.

Exercise 8—Factors That Affect Enzyme Activity

1.   We've seen that unusually high temperatures can affect the three-dimensional folding of a protein. In enzymes, if the structure of the active site is altered, the ability of the enzymes to function may be destroyed. High fevers may affect enzymes throughout the body.

2.   Enzymes are sensitive to changes in environmental conditions, and many are quite specific as to the pH range in which they can function.

3.   pH 3.5

4.   At pH 3.7, it took 14 minutes before noticeable enzyme activity was observed. In comparison, enzyme activity was observed after only 2 minutes at pH 3.5.

5.   The enzyme is specific because its optimal function was centered on pH 3.5. Enzyme activity at pH values even slightly above or below 3.5 was greatly decreased.

6.   The production of melanin (and, thus, the fur color in Siamese cats and Himalayan rabbits) is related to body temperature. This trait does not apply to the colors of skin, fur, and feathers in all animals, just these specific breeds. Both blood circulation and body temperature decrease in the extremities. Because body temperature is lower in the paws, ear tips, and nose, the enzyme for melanin production is active. Core temperatures at the center of the body are too high for melanin synthesis to be completed.

Exercise 9—Introduction to Molecular Genetics

1.   A—DNA

2.   C—both DNA and RNA

3.   A—DNA

4.   A—DNA

5.   D—neither DNA nor RNA

6.   B—RNA

7.   C—both DNA and RNA

8.   B—RNA

9.   A—DNA

10.   C—both DNA and RNA

11.   Yes, your friend has misunderstood. You wouldn't be able to tell this from circulating red blood cells because they contain no nucleus and, therefore, no DNA.

12.   A chromosome consists of a long strand of DNA wrapped around proteins. This long strand of DNA is divided into units called genes. Genes contain the coded instructions for the synthesis of proteins. Proteins are made of linked amino acids. The coded instructions are written in the form of three-letter "words" called triplets. Each triplet consists of three DNA bases. The sequence of DNA triplets specifies the number of amino acids in the protein, which amino acids are needed, and the sequence in which the amino acids must be connected to produce the correct primary structure.

To initiate the process of protein synthesis, a complementary copy of the DNA instructions (messenger RNA) is made. The mRNA copy also contains three-letter code words that are referred to as codons. The mRNA leaves the nucleus and is picked up by a ribosome in the cytoplasm for the process of protein synthesis to continue.

13.   6

14.   C A A

15.   G U U

16.   amino acid #1—isoleucine

amino acid #2—glycine

amino acid #3—leucine

amino acid #4—serine

17.   No, because there is more than one codon that codes the same amino acid. In this particular case, both CCT and CCC code for glycine. Other mutations could cause the substitution of an incorrect amino acid in the protein, but not this one.

Exercise 10—Mitosis and Asexual Reproduction

1. E—nuclear membrane

2. I—equator

3. C—centromere

4. K—daughter cells

5. A—diploid

6. D—spindle fibers

7. H—cleavage furrow

8. B—sister chromatid

9. G—cytokinesis

10. L—chromosome

11. False

12. True

13. True

14. True

15. Stages of mitosis for parent cell with four chromosomes:

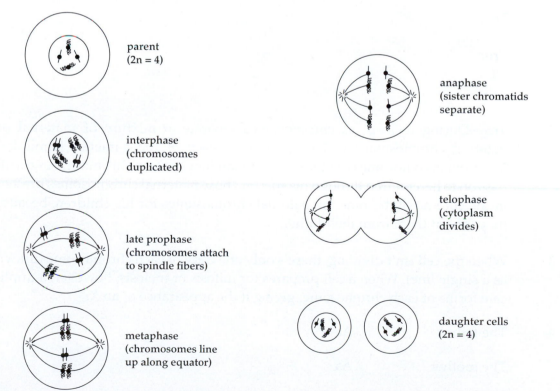

parent
(2n = 4)

interphase
(chromosomes
duplicated)

late prophase
(chromosomes attach
to spindle fibers)

metaphase
(chromosomes line
up along equator)

anaphase
(sister chromatids
separate)

telophase
(cytoplasm
divides)

daughter cells
(2n = 4)

16.   The plan will not be effective. The fishermen are actually doubling the number of starfish. Starfish are capable of regeneration, which is an example of asexual cell division. Each half of the starfish can replace lost body tissues and develop into a new, fully functional adult.

17.   A—Skin cells are diploid and have a full set of chromosomes necessary to provide the genetic information to make a new body. The other two options are not viable because sperm cells are haploid and red blood cells have no nucleus.

Exercise 11—Connecting Meiosis and Genetics

1.   False

      True

      False

      False

      False

      True

      False

      True

      True

      False

2.   Yes—During meiosis, as chromosomes separate, a mixture of maternal or paternal chromosomes will be present in each of the gametes. Reminder: Maternal chromosomes are those that a man inherits from his mother and will pass on to his children through his sperm. Those maternal chromosomes present in the sperm will become the paternal chromosomes for his children, because they inherit them from their father.

3.   When the cell isn't dividing, there's only one copy of each chromosome (drawn as a single line). When a cell prepares for mitosis or meiosis, however, a duplicate forms of each chromosome, giving it the appearance of an X.

4.   The father   Aa

      The mother   Aa

The child	aa
Probability of an albino child	25%
Probability of a normally pigmented child	75%

Explanation: A child inherits one allele from each parent; therefore, each parent must have one little "a."

5.

Mother	Tt
Father	Tt
First child	tt

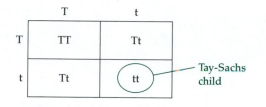

Explanation: A child inherits one allele from each parent; therefore, each parent must have one little "t." The parents could not, however, be homozygous recessive because people with Tay-Sachs disease die in early childhood.

6.

Ghandi	Cc
Sabrina	Cc
Snowflake	cc
Probability of heterozygous cub	50%

7.

Ghandi	Cc
White tiger at other zoo	cc
Probability of another white cub	50%

Exercise 12—Human Genetics

1.

The woman	tt
Her husband	Tt

The two sons	Tt
The daughter	tt
The woman's parents	Tt and Tt
The husband's parents	T? and T?

2. | Your genotype | Dd |
Your husband	DD
Your sister	dd
Probability of Tay-Sachs child	0%

3. | Fred | X^BY^0 |
Ginger	X^BX^b
David	X^bY^0
Takiyah	$X^BX^?$ but probably X^BX^B
Kelly	X^BX^b
Kevin	X^bY^0
Takiyah's five sons	X^BY^0
Probability of color-blind son	25%
Probability of color-blind daughter	0%

4. | Ralph | X^HY^0 |
Ralph's brothers	X^hY^0
Ralph's sister	X^hX^h
Ralph's mother	X^HX^h
Ralph's father	X^hY^0

Because Ralph has normal blood clotting, he must have inherited a big "H" from his mother. Because Ralph has a hemophiliac sister, however, his mother must be heterozygous. A child inherits one allele from each parent, so to produce a hemophiliac daughter, both parents must carry the "h" allele. Therefore, Ralph's father must be a hemophiliac.

5.   Pedigree showing albino individuals:

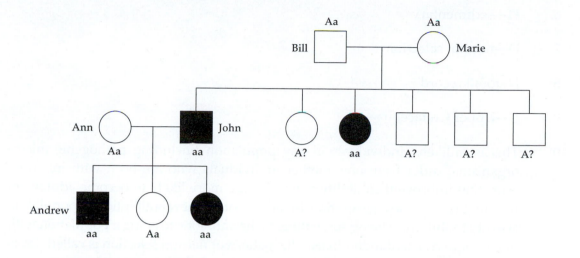

Albinism is inherited through a recessive allele. Neither Bill nor Marie is albino, yet they had an albino son, John. In a dominant/recessive trait, this is only possible if both parents carried (but didn't express) a recessive allele for the trait. An albino individual would be homozygous recessive, and it wouldn't be possible for two albino parents to have a child with normal skin color because neither parent possesses an allele for normal pigmentation to pass on to his or her offspring.

6.   Based on the results of a blood test, it isn't possible to conclude that Neil and Alice have the wrong baby. Because Neil has type A blood, his genotype can be either $I^A I^A$ or $I^A i$.

Alice has type B blood, which means her genotype is either $I^B I^B$ or $I^B i$. The baby has type O blood, which has only one possible genotype, ii. Also, the baby has to inherit one allele from each parent. If Neil's genotype was $I^A i$ and Alice's genotype was $I^B i$, the baby could have inherited an "i" from each parent and have type O blood.

To be completely sure that the baby is theirs, DNA testing would be required.

Exercise 13—Evolution

1.   K—radiometric dating

2.   A—Charles Lyell

3.   I—isotope

4.   E—evolution

5.   B—Charles Darwin

6.   H—sedimentary

7.   D—natural selection

8.   J—fossil record

9.   L—Upper Cretaceous

10.   This is unlikely. Individuals in any population (including pathogenic micro-organisms) differ from one another. Individuals who are born with inherited traits that improve their ability to survive are more likely to become adults and the parents of the next generation. In the case of this new antibiotic, the population is likely to shift from being susceptible to the antibiotic to being a population with many more resistant individuals. This pattern of natural selection is called directional selection, because the allele frequency has shifted in a specific direction.

11.   The half-life of the isotope is 700 million years. That means that every 700 million years, half of the original amount of ^{235}U will have been converted to another element. At time zero, we had 20 mg of the isotope. Seven hundred million years later, we would have only 10 mg remaining and after another 700 million years had passed, we would have 5 mg (the amount discovered in your dig). Therefore, you're digging in a rock stratum that's 1.4 billion (1,400 million) years old.

12.   Individuals in any population (plants, animals, microorganisms) differ from one another, and many of these differences are genetically based. This principle is referred to as genetic variability. Individuals who are born with inherited traits that improve their ability to survive are more likely to become adults and the parents of the next generation. They pass these inherited "survival" traits on to many of their offspring. This causes a shift in the gene frequency of the next generation, a process defined as evolution. Over several generations, more members of the population will share those beneficial traits.

Exercise 14—Functions of Tissues and Organs

1.   I—sebaceous gland

2.   J—dermis

3.   A—keratin

4.   D—epithelial cells

5.   C—erector muscle

6.   H—subcutaneous layer

7.   B—melanin

8.   E—epidermis

9.   The dermis. The nicotine medication must penetrate through the epidermis and enter blood vessels in the dermis.

10.   On the sole of the foot, the epidermal layer is thickened and calluses may be present. Penetration of the patch medication through to the dermis might prove difficult.

11.   Your friend is experiencing muscle fatigue. The actively contracting muscles become weaker as time passes. This can be caused by lack of ATP, insufficient oxygen, depletion of energy reserves in the muscle cells, and accumulation of metabolic wastes.

12.   In second-degree burns, the epidermis has been destroyed, exposing the sensitive nerve endings in the dermis. Pain will be severe. In third-degree burns, nerve endings are completely destroyed. Consequently, the burn patient feels no pain until the nerves begin to regenerate and repair themselves.

Exercise 15—The Cardiovascular System

1.   1—tissue capillaries in big toe

        6—pulmonary artery

        9—left atrium

        8—pulmonary vein

        4—right atrium

        10—left ventricle

        3—inferior vena cava

        11—aorta

        5—right ventricle

        12—arterioles

        2—venules

        7—lungs

2.   O—tissue capillaries entering big toe

        D—pulmonary artery

O—left atrium

O—pulmonary vein

D—right atrium

O—left ventricle

D—inferior vena cava

O—aorta

D—right ventricle

D—tissue capillaries leaving big toe

O—arterioles

D—venules

3.   Blood would back up into the right atrium.

4.   96

5.   Owing to rheumatic fever, scarred heart valves are not closing completely. The hissing sound is caused by a small amount of blood leaking through the valve under pressure when the ventricles contract.

6.   Skin arteries dilate (enlarge) to increase blood flow to the skin.

7.   Right ventricle—40   pulmonary artery—40

     Pulmonary vein—100   left atrium—100

8.   Blood flow to the muscles increases from about 21% to about 73%. Diameter of blood vessels supplying muscle tissue also increases.

9.   Blood flow to the abdominal organs decreases from about 25% to about 3%. Diameter of blood vessels supplying abdominal organs also decreases.

10. Blood flow to the skin increases from about 9% to about 12%. Increased blood flow to the skin is one mechanism by which the body releases excess heat during exercise.

Exercise 16—Introduction to Anatomy: Dissecting the Fetal Pig

1.   C—head end of the body

2.   A—back

3.   B—tail end of the body

4.   D—belly side

5.   Female pigs have a genital papilla just ventral to the anus. In addition, the urogenital opening of females is located just below the papilla (as opposed to males, in which the urogenital opening is located just posterior to the umbilical cord).

6.   D—The trachea is the only air passageway to the lungs. Although the esophagus is also located in the throat, it is not involved in the breathing process.

7.   Esophagus

8.   Diaphragm

9.   Under heavy exercise, a person with emphysema will not get enough oxygen for the body's needs. The maximum amount of air moved in and out (vital capacity) when breathing deeply is significantly reduced.

10.   When insufficient oxygen is transported by red blood cells, the mitochondria in body cells cannot function at peak efficiency, and the rate of cell respiration decreases. The amount of ATP produced decreases and, therefore, the amount of usable energy available to power cell activities will also decrease.

Exercise 17—Organs of the Abdominal Cavity

1.   H—mesenteries

2.   C—villi

3.   D—circular muscle

4.   F—peristalsis

5.   E—longitudinal muscle

6.   B—root hairs

7.   Digestive system: stomach, small intestine, large intestine, pancreas, liver, and gall bladder

        Respiratory system: no organs from this system in the abdominal cavity

        Urinary system: kidney, bladder, and ureter

        Circulatory system: arteries, veins, and capillaries

8.   Cardiac sphincter; stomach acids back up and irritate the lining of the esophagus. Because this region of the esophagus is not far from the heart, this discomfort can be mistaken for chest pains.

9.   Stomach acids react with the enamel on the teeth and gradually dissolve the enamel over long periods of time.

10. The malfunctioning organ is probably the bladder. The malfunctioning structure is probably the sphincter associated with the bladder and urethra, which prevents urine from being released accidentally.

11. Design B will be more efficient because more interior surface area is in contact with the air.

12. Crossword puzzle answers:

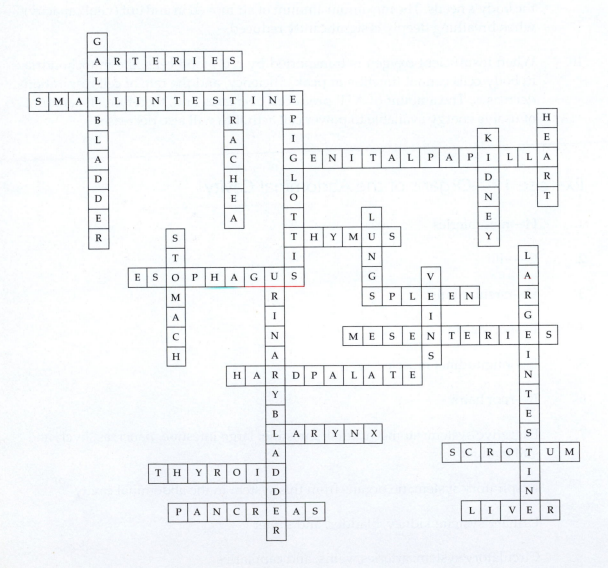

Exercise 18—Introduction to Forensic Biology

1.   C—whorl

2.   G—divergence

3.   F—bifurcation

4.   E—loop

5.   J—sweat glands

6.   H—fingerprint formula

7.   I—points of similarity

8.   D—tented arch

9.   The skin is covered with sweat glands that produce perspiration, which can accumulate on the ridges, forming the fingerprints. These ridges also accumulate body oils from touching oily surfaces. Together, these leave an invisible impression, which is the fingerprint.

10.   Even though their fingerprints might be the same at birth, differing daily activities cause an accumulation of scars and other marks that can be used to tell the fingerprints of the twins apart.

11.   The blood type is A–. Agglutination with anti-A serum shows the presence of type "A" proteins, but lack of agglutination with anti-B and anti-Rh serums shows that these proteins are absent.

12.   The blood type is AB–. Agglutination occurred with both anti-A and anti-B serums, showing the presence of type "A" and type "B" proteins, but lack of agglutination with anti-Rh serums shows that this protein is absent.

13.   The blood type is O+. Agglutination did not occur with either anti-A or anti-B serums, showing the absence of type "A" and type "B" proteins, but agglutination did occur with anti-Rh serum, which shows that the Rh protein is present.

Exercise 19—Biotechnology: DNA Analysis

1.   Noncoding DNA is that portion of the DNA molecule that doesn't code for the synthesis of proteins. Although the functions of these areas are not yet clearly defined, scientists suspect that they play an important role in RNA synthesis and other cellular functions.

2.   The pattern of repeating noncoding DNA sequences differs from person to person. This gives each person a unique DNA profile through the process of RFLP analysis.

3.   No child is genetically identical to either of the parents. Through the process of meiosis and fertilization, each child inherits half the genes from each parent. By analyzing the child's STR profile and comparing it with that of both parents, it's possible to determine which bands were not inherited from the mother and therefore to conclude that they must have been inherited from the father. In blood sample comparisons, however, the entire STR profile must be a match between the two samples to show that both blood samples came from the same person.

4.   The deer meat could be tested by STR analysis. If the deer's STR profile matched that of the skin or internal organs found inside the park, this would demonstrate conclusively that the confiscated deer was the individual killed within the park boundaries.

5.   The blood found on the dog's collar matches the owner's blood. An analysis of the STR profile of the child's blood does not match that of the sample from the dog's collar. This evidence substantiates the owner's claim that her dog was not responsible for the attack.

Exercise 20—Using Biotechnology to Assess Ecosystem Damage

1.   Product A was most toxic at full strength because the bacteria glowed only at an intensity level of one. Product A was determined to be harmless at 1:100 because this was the first dilution at which bacteria glowed at an intensity level of four.

2.   There was no dilution level at which Product B was harmless. Even at the lowest dilution (1:10,000), the bacteria glowed only at an intensity level of two.

3.   Product A is least harmful to the environment because it is less toxic at lower dilution levels. It can be safely disposed of by diluting with water in the ratio of 1:100.

Exercise 21—Ecosystems

1.   Examples: pine trees, rose bushes, corn, rice, beans, photosynthetic bacteria (any organism that contains chlorophyll is acceptable). In addition, not all producers perform photosynthesis. Others perform chemosynthesis—the conversion of minerals to usable energy. Most of these organisms are bacteria.

2.   weeds   →   crickets   →   rats   →   cats

        producer   →   primary   →   secondary   →   tertiary
                            consumer   consumer   consumer

3.   If one of the cats died of natural causes in the vacant lot, it would be consumed
        by decomposer organisms. In this way, nutrients contained in the cat's body
        would be recycled and the decomposers would obtain energy.

4.   Crossword puzzle answers:

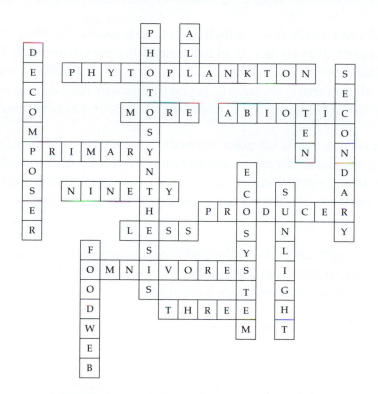

Exercise 22—Population Ecology

1.   C—too high

2.   B—mark and recapture

3.   E—95% confidence interval

4.   D—too low

5.   Accuracy would be affected if the trees had a patchy (rather than even)
        distribution in the forest. In this situation, your sampled quadrats might not
        be truly representative of the tree population and your estimate would be too
        low. One cause of patchy distribution would be a change in the environmental

conditions from one place to another in the study area (for example, if one area has a higher elevation or more water). If one area was exposed to forest fire or a disease, adult pines might be less numerous in this area, affecting your sampling.

6. In order to develop an effective management plan and set catch limits for a commercially important species of fishes, ecologists need a lot of information about the species. For example, what is the percentage of individuals of various ages in the population? Is there seasonal migration? What time of year do female fish spawn? (You wouldn't want to catch any during this period.) This and other life history information can be obtained from mark and recapture data.

7. What you're actually doing with this method is measuring your catch per unit of fishing effort. This gives you an estimate of the population size. If you put in a lot of effort (many hours of fishing) but your catch is relatively small, you know the population size is decreasing. If you catch many tuna during your 100 hours, you can feel confident that the population size is at a healthy level.

8. Some species are difficult to see when they are active, and many frog species are active in the dark. Therefore, vocalization frequency is an established method of estimating population size in this type of situation. The same method is also used to estimate population size of birds and insects that make sounds (crickets, cicadas).

9. You could mark and release the cockroaches caught in the trap. Then calculate the population size from the number of marked individuals recaptured as you did in your mark and recapture experiment.

Photo/Illustration Credits

Page 301: Figure 17-4a: David Musher/Photo Researchers, Inc.

Page 301: Figure 17-4b: Education Resources

Page 306: Figure 17-5: Education Resources

Page 340: Figure 19-1: Courtesy of Orchid Cellmark, Inc.

Page 398: Figure 22-1: Dreamstime LLC

Page 399: Figure 22-2: Peter P. Panyon Jr., Prince George's Community College

Page 407: Figure 22-5: A Hartl/age fotostock

Pages 417–422: Environmental Jenga game: Adapted from Whale Jenga by Cabrillo Marine Aquarium (cabrillomarineaquarium.org)